Performance Interventions

Series Editors: **Elaine Aston**, University
University of California, Irvine

Performance Interventions is a series of monographs and essay collections on theatre, performance, and visual culture that share an underlying commitment to the radical and political potential of the arts in our contemporary moment, or give consideration to performance and to visual culture from the past deemed crucial to a social and political present. *Performance Interventions* moves transversally across artistic and ideological boundaries to publish work that promotes dialogue between practitioners and academics, and interactions between performance communities, educational institutions, and academic disciplines.

Titles include:

Alan Ackerman and Martin Puchner (*editors*)
AGAINST THEATRE
Creative Destructions on the Modernist Stage

Elaine Aston and Geraldine Harris (*editors*)
FEMINIST FUTURES?
Theatre, Performance, Theory

Maaike Bleeker
VISUALITY IN THE THEATRE
The Locus of Looking

Clare Finburgh and Carl Lavery (*editors*)
CONTEMPORARY FRENCH THEATRE AND PERFORMANCE

James Frieze
NAMING THEATRE
Demonstrative Diagnosis in Performance

Lynette Goddard
STAGING BLACK FEMINISMS
Identity, Politics, Performance

Alison Forsyth and Chris Megson (*editors*)
GET REAL: DOCUMENTARY THEATRE PAST AND PRESENT

Leslie Hill and Helen Paris (*editors*)
PERFORMANCE AND PLACE

D.J. Hopkins, Shelley Orr and Kim Solga
PERFORMANCE AND THE CITY

Amelia Howe Kritzer
POLITICAL THEATRE IN POST-THATCHER BRITAIN
New Writing: 1995–2005

Marcela Kostihová
SHAKESPEARE IN TRANSITION
Political Appropriations in the Post-Communist Czech Republic

Jon McKenzie, Heike Roms and C.J W.-L. Wee (*editors*)
CONTESTING PERFORMANCE
Emerging Sites of Research

Jennifer Parker-Starbuck
CYBORG THEATRE
Corporeal/Technological Intersections in Multimedia Performance

Ramón H. Rivera-Servera and Harvey Young
PERFORMANCE IN THE BORDERLANDS

Mike Sell (*editor*)
AVANT-GARDE PERFORMANCE AND MATERIAL EXCHANGE
Vectors of the Radical

Melissa Sihra (*editor*)
WOMEN IN IRISH DRAMA
A Century of Authorship and Representation

Brian Singleton
MASCULINITIES AND THE CONTEMPORARY IRISH THEATRE

Performance Interventions
Series Standing Order ISBN 978–1–4039–4443–6 Hardback 978–1–4039–4444–3 Paperback
(*outside North America only*)

You can receive future titles in this series as they are published by placing a standing order. Please contact your bookseller or, in case of difficulty, write to us at the address below with your name and address, the title of the series and the ISBN quoted above.

Customer Services Department, Macmillan Distribution Ltd, Houndmills, Basingstoke, Hampshire RG21 6XS, England

Visuality in the Theatre

The Locus of Looking

Maaike Bleeker

palgrave
macmillan

First published in hardback 2008
First published in paperback in 2011 by
PALGRAVE MACMILLAN

Palgrave Macmillan in the UK is an imprint of Macmillan Publishers Limited, registered in England, company number 785998, of Houndmills, Basingstoke, Hampshire RG21 6XS.

Palgrave Macmillan in the US is a division of St Martin's Press LLC, 175 Fifth Avenue, New York, NY 10010.

Palgrave Macmillan is the global academic imprint of the above companies and has companies and representatives throughout the world.

Palgrave® and Macmillan® are registered trademarks in the United States, the United Kingdom, Europe and other countries.

ISBN 978–0–230–54709–4 hardback
ISBN 978–0–230–30084–2 paperback

This book is printed on paper suitable for recycling and made from fully managed and sustained forest sources. Logging, pulping and manufacturing processes are expected to conform to the environmental regulations of the country of origin.

A catalogue record for this book is available from the British Library.

Library of Congress Cataloging-in-Publication Data

Bleeker, Maaike.
Visuality in the theatre : the locus of looking / Maaike Bleeker.
p. cm.—(Performance interventions)
Includes bibliographical references and index.
ISBN 978–0–230–54709–4 (cloth) 978–0–230–30084–2 (pbk)
1. Theater—Philosophy. I. Title.

PN2039.B59 2008
792.01—dc22 2008011741

10 9 8 7 6 5 4 3 2 1
20 19 18 17 16 15 14 13 12 11

For my sister Bregje

Contents

List of Illustrations

Acknowledgements

In *What is Philosophy?*, Deleuze and Guattari conceive of philosophy in terms of friendship. This philosophical friendship, as Deleuze explains, is not based on having the same ideas. Rather, it is the condition of having something to say to another as a result of which thought starts to move. Friendship carries something of a mystery within it, and this mystery Deleuze understands in terms of being possessed by a certain 'charme' that lights between friends, turning them into friends.

During the years of research and writing that have resulted in this book, many friendships – philosophical and other – have lighted my way. These are too many to mention. I will mention some. In the ASCA community in Amsterdam, I found a first audience for many of my texts and ideas; in research centre Aisthesis in Antwerp and the Choreography and Corporeality working group of the International Federation for Theatre Research a second and a third. Of these many dear friends, there are two that I would like to thank in particular. These are Mieke Bal and Susan Foster. Mieke presented me with an invaluable example of how to do things within the academy, and outside of it: of being personal, passionate and precise. She taught me how to do things with theory. Susan Foster made me dance with it. Embodying the Deleuzian idea of philosophy in terms of movement and friendship, she provided the spark that challenged me to twist and turn, to take off and jump.

I would like to thank all the artists whose work is discussed on the pages of this book for their thought provoking aesthetic texts as well as for the interest in my project that many of you have shown. Thank you and the photographers who documented your work for the kind permission to reproduce some of your work alongside my text. I also want to thank those who have given me the opportunity to share some parts of my work in progress with a wider audience. Some smaller parts of Chapters 2, 8 and 9 have been part of articles which appeared in Performance Research ('Death, Digitalization and Dys-appearance. Staging the Body of Science', *Performance Research*, Vol. 4, no. 2 (1999) 1–7; 'Being Where? Managing the Attention of the Audience, in Beppie Blankert's *Double Track*', *Performance Research*, Vol. 6, no. 3 (2001) 104–110;

'Absorption and Focalisation: Performance and it's Double', *Performance Research*, Vol. 10, no. 1 (2005) 48–60). Some parts of Chapter 3 have been part of an article which appeared in Theatre Research International ('Look Who is Looking! Perspective and the Paradox of Post-Dramatic Subjectivity', *Theatre Research International*, Vol. 29, no. 1 (2004) 29–41). An earlier version of Chapter 6 was published in Anke Bangma (ed.) *Looking, Encountering, Staging* (Rotterdam: Piet Zwart Institute in collaboration with Revolver Archiv für Actuelle Kunst, 2005) pp. 27–42, and is reprinted here with permission of the editor. An earlier version of Chapter 7 was published in Maaike Bleeker, Steven Debelder, Kaat Deboo, Luk van den Dries, Kurt Vanhoutte (eds) *Body Check. Relocating the Body in Contemporary Theatre* (Amsterdam: Rodopi, 2002), pp. 131–160, and is reprinted here with permission of Rodopi publishers. And thank you Bryan Reynolds for your confidence in this book project and your enthusiastic support.

With my sister Bregje, I share a history that spans almost my entire life, and all of hers. This history has presented us with many questions but, luckily, also with each other to share the quest for answers. In search of answers, we typically choose opposite directions. Nevertheless, we both end up writing books. This book is my way of dealing with this history we share and some of the questions it raises. In admiration of your ways of dealing with these very same questions, and hoping for many more years of writing together, my dearest Bregje, I dedicate this book to you.

Last but not least, thank you Maurice, my shining star, for being so much more than a friend.

1
Dissecting Visuality in the Theatre – Introduction

In 1994, I attended Peter Brook's *The Man Who*, in which 4 actors embody 13 cases of neurological disorder.[1] I was drawn into the world of persons suffering from autism, Korsakoff's syndrome, amnesia, epileptic memory, Tourette's syndrome and many other ailments. In presenting these diseases as forms of embodiment that are subjectively profound and symbolically significant (rather than statistically measurable external attacks on the biochemical organism), *The Man Who* does justice to the brilliant medical studies of Oliver Sacks on which it is based.[2] Furthermore, in demonstrating unexpected relationships between particular variations of embodiment and ways of perceiving the world, *The Man Who* tackles some remarkable implications of the 'located-ness' of the act of looking from within a particular body. As performed in *The Man Who*, these cases present the body as the very medium though which a vision of the world comes into being. The theatrical exposition of these neurological cases, therefore, does not simply offer a demonstration of how things are (with these patients), it contains an argument as well. *The Man Who* reads as a critique of the strict division between vision, the body and the other senses typical of the modern understanding of self, of the world, and of the theatre. For this reason, *The Man Who* is exemplary for both the subject of this study – dissecting visuality – and the object of my analysis – the theatre.

A new or renewed focus on questions of vision in a wide variety of fields has begun to open our eyes to the complexity of what easily, but mistakenly, is taken for granted as 'just looking'. Growing awareness of the inevitable entanglement of vision with what is called *visuality* – the distinct historical manifestations of visual experience – draws attention to the necessity of locating vision within a specific historical and cultural situation. This is a situation in which what we think we see is the product of vision 'taking place' according to the tacit rules of a specific scopic regime and within a relationship between the one seeing and what is seen.[3] What seems to be just 'there to be seen' is, in fact, rerouted through memory and fantasy, caught up in threads of the unconscious and entangled with the passions. Vision, far from being the 'noblest of the senses' (Descartes, 1977), appears to be irrational, inconsistent and undependable. More than that, seeing appears to alter the thing seen and to transform the one seeing, showing them to be profoundly intertwined in the event that is visuality. 'There is no such thing as just looking', as James Elkins puts it (Elkins, 1996, p. 31).

Visuality happens. Visuality is not a given property of things, situations, or objects. Visuality is not even an object in the sense that film is the object of film studies or the work of art the object of art history. The study of visuality draws upon work accomplished within many different disciplines and engages with images, texts, or events that have been studied and still are studied within these different disciplines, yet its own object domain – visuality – does not belong to any of these disciplines proper. Instead, the object of visual analysis is the way things become visible as a result of the practices of looking invested in them.[4] Visuality as an object of study, therefore, requires that we focus on the relationship between the one seeing and what is seen.

This relationship between someone seeing and what is seen is often considered to be a fundamental characteristic of the theatrical event and crucial to the intense experiences it can evoke. The word theatre is derived from the Greek *theatron* which denotes the place from where the theatrical event is seen. The theatre organizes the relation between those seeing and what they see, mediating in a specific relationship between the two. The theatre, therefore, (or so it would seem) presents the object *par exellence* for an analysis of visuality as a phenomenon that takes place within the relationship between the one seeing and what is seen and against the backdrop of culturally and historically specific visual practices. All the more remarkable is the almost total absence of references to the theatre in interdisciplinary textbooks and studies on visuality, and vice versa the apparent lack of interest in visual analysis from the perspective of theatre studies.

Some point to the tradition of staging texts that for a long time dominated the history of the Western theatre. In this tradition, all the other elements of theatrical performance were understood to be merely supportive of this text, and therefore often ignored. However, this explanation ignores the fact that text-based theatre too has many visual aspects. Furthermore, this explanation can neither account for the ways in which text actually plays an important part in vision as it 'takes place' in the theatre, nor can it explain *how* the visual aspects of text-based theatre can appear as merely supportive to the text. The word–image opposition that is evoked obscures the fact that (especially in text-based theatre) text is instrumental in how theatre invites ways of looking and mediates in a particular relation between the one seeing and what is seen.[5] Finally, what such an explanation fails to acknowledge is that the opposition of text-based theatre to more visually oriented theatre is itself a product of the scopic regime of modernity and is itself supportive of its object immanent (rather than relational) understanding of visuality.

Could it be that the very obviousness of the theatre as a means of staging something for a viewer itself obscures the relational character of visuality as it 'takes place' in the theatre? The staged character of the theatrical event invites an understanding of what we see as the result of what is being done for us, a situation in which we as audience are 'just looking'. Indications of our (the viewer) being implicated in what is seen can easily appear as merely the product of what 'they' are doing for us, without taking into consideration our own involvement in how we see what we see. Such moments when we do become aware of our own being implicated in what is seen, tend to get marked as 'theatrical': what we see appears as inauthentic or false, and the address presented to us is dismissed as exaggeration or make believe. Typically, the adjective *theatrical* refers both to a particular quality of something – its being 'of the theatre' – and to failure: the failure to convince as authentic and true. Typically, from Diderot to deconstruction, the history of staging the relationship between the one seeing and what is seen in the theatre is marked by *anti*-theatricality. The staged character of the theatrical event makes it by definition antithetic to modernist notions of authenticity and truth, thus condemning the theatre to presentational strategies that, in order to convince as true and authentic, have to be aimed at obscuring or erasing traces of its own condition of being staged. When this relational character of the theatrical event does become visible, it can only appear as failure.

This complexity can be seen reflected in many accounts of the theatre in which the relationship between audience and event is described in terms of immediacy and directness, that is, in terms that, in describing this relationship, at the same time deny this relationship qua relationality. Take for example these accounts of the contemporary stage where, breaking theatre conventions, performers offer us glimpses (or more) of 'themselves' as living, breathing beings, standing literally or figuratively naked before our very eyes. One is able to see all of them, to see right

Not ST

through them, and to feel what they feel, sensing the physical reactions of their bodies seen on stage as though they were one's own, or so it seems. The theoretical attention paid to these bodies seen on stage produces a body of knowledge that testifies to growing awareness of the way bodies are involved in meaning production on stage and of the way these bodies seen are the products of culture rather than natural givens. At the same time, this body of knowledge bears witness to a rather disembodied notion of what it means to see these bodies. It is disembodied notion of vision that allows for the conflation to take place between what is seen and what is present 'over there'. This disembodied notion of vision supports the tendency to take what is seen for what is 'over there', and to understand the strong effects thus experienced as resulting exclusively from the body present on stage. What is left out, then, is the relation between the body seen and the body seeing.

Could it be that the relationality of vision as it 'takes place' in the theatre is so hard to see because as theatricality it is the repressed other of the modern visual paradigm, constitutive to its construction but precisely as that which has to remain invisible? If so, the theatre might be called paradigmatic for what Jonathan Crary (1999) describes as practices of *managing attention* typically of the late modern period. Attention emerges as a discursive and practical object right at the moment that vision and hearing had become progressively severed from the various codes and practices that had invested them with a level of certainty, dependability, and naturalness. Crary locates its appearance early in the 19th century, at the moment that the Cartesian notion of objective, disembodied vision begins to loose its self-evident character. At this historical moment, attention emerges both as a problem and as the promise of a solution to this very same problem. If vision can no longer automatically be associated with truth and objectivity, this means we can be misled by what we see, either because of our own subjective limitations, or because of purposeful deceit. However, strategies of managing attention now appear as a promise of the possibility of transcending these selfsame subjective limitations and therefore as a means to safeguard the possibility of objective and true vision against the threat of subjectivity. Crary refers to the rise of phenomenology and its method of 'bracketing' as well as to the ways in which in much later-19th- and early-20th-century aesthetic theory posited various modalities of contemplation and vision that were radically cut off from the processes and activities of the body. One of his examples of historical strategies of managing attention is, surprisingly, a theatre: Wagner's *Festspielhaus* in Bayreuth (1876). Phenomenology, formalism as well as the architecture of the *Festspielhaus* appear as attempts at restoring or preserving the autonomous self-knowing and unitary subject of Cartesian thought. They mediate in ways of seeing that typify the disembodied and detached eye/I right at the moment that this same eye/I came under attack by subjective and embodied notions of vision. They do so by means of strategies of 'staging' the relation between the one seeing and what is seen in such a way as to obscure precisely the relationship between what is seen and the subjective point of view from where it is seen as such.

Crary's analysis turns the theatre into the paradigmatic example of a wide variety of cultural practices all aiming at maintaining and preserving the modern fable of vision as true and objective, of the possibility of seeing it 'as it is'. The theatre presents a model for understanding these practices as ways of organizing the relationship between the one seeing and what is seen, while at the same time theatricality (and by extension the theatre itself) is what has to be repressed in order to safeguard the illusion of the seen as evidence, as truth and fact; of the apparent autonomy of distance and separateness to the spectator, and of the visual aspects of the world seen as being properties of this world itself. This illusion is the primary object of my 'dissection of visuality'.

The anatomy of 'just looking'

Looking back at *The Man Who*, I remember one scene in particular. In this scene, three patients have lost awareness of the left side of their bodies and, as is shown, this seriously affects their perception of the world around them. The first patient is convinced that his left hand is actually his mother's. The second mistakes the hand of the doctor for his own and perceives only the right side of whatever is before him. The third, asked to describe the *Place de la Concorde* in Paris as he sees it in his imagination, gives a complete description of the right side of the square only. When he is asked to 'turn around' and imagine the square from the opposite direction, again he seems to be aware of the right side only and now the images of the left side have disappeared completely. He is not able to recall them or to relate descriptions given by the doctor to his mental image of the square. Later on, he attempts to get rid of his left arm and leg. To him, these bodily parts seem to be part of a dead body put into his bed.

It strikes me that the loss of awareness of a part of their body not only prevents these people from perceiving stimuli with the left side of their body, but radically alters their perception of the world around them as well. These patients are not paralysed or disabled. Rather, it appears that their bodily schemata have been altered in such a way that they do no longer include the left half of their physical presence. This results in remarkable changes in what they consider to be 'self' as opposed to 'other'. These changes contradict the conception of the physical boundaries of the body-organism as the natural and necessary boundaries of the embodied subject. Furthermore, these alterations in their bodily schemata not only affect the way these patients become aware of their own bodily or physical presence, but appear to change their way of looking as well, and in this way altering their understanding of the world surrounding them. In doing so, Brook's play presents what seems to be an 'inverse' of the relationship between bodies, space, vision and subjectivity as it is part of the psychoanalytical story of the subject, the modern story of the subject *par exellence*.

Freud acknowledges that our experience of self is always circumscribed by and derived from the body. He claims that the ego's outline is a psychic map or projection of the surface of the body that provides the basis of the subject's assumption that it is coextensive with the whole of its body. This idea of taking up the body as its body was further developed by Lacan, who uses the opposition between tactile and kinesthetic information (which yields the fragmented image of the body in bits and pieces) and visual perception (which provides the illusionary unity of the image as an ideal image or mirror for the subject) to explain the genesis of an always alienated ego. In his famous essay on the mirror stage, Lacan points to the close relationship between the bodily ego and the field of vision. Lacan writes that 'the mirror image would seem to be the threshold of the visible world' and it is only by moving through the mirror stage that one enters the scopic domain (Lacan, 1977, p. 3).[6] What these stories by Freud and Lacan have in common, is that in both, as Elizabeth Grosz remarks: 'the ego forms itself around a fantasy of a totalized and mastered body, which is precisely the Cartesian fantasy modern philosophy has inherited' (Grosz, 1995, p. 86). The ego takes up its body as unity seen from a point of view outside the body, and by means of a seeing that is separated from – even opposed to – the body and the other senses. Grosz therefore proposes a move analogous to the one suggested by Brook's play, namely to invert the primacy of a psychical interiority by demonstrating its necessary dependence on a corporeal exteriority, and to understand this exteriority as the very 'stuff of subjectivity'. The possibilities of such a reconfiguration of the subjectivity involved in visual perception and the implications for our understanding of the spectator as body perceiving (rather than a disembodied eye/I) is what my dissection of visuality will work towards.

Performing analysis

I will approach the body/subject of perception through a dissection of the discursive determinations that produce seeing as the 'just looking' of the disembodied eye/I. To dissect is to cut into pieces, to examine part by part, to analyse and to criticize in detail. In its medical sense, dissection is the methodical division of a human or animal body to show its parts and structure and to investigate its morphology. Once one method among many others, dissection has become the model of scientific investigation as a practice concerned with producing objective visions of the world 'as it is'. Visions, that is, cut loose from the subjective point of view of a particular observer.[7]

With dissecting visuality in the theatre, I propose a deconstructive reversal of this movement. The object under dissection here is not some body but visuality. My 'dissection' does not aim at making this visuality visible as what it is 'in itself', independent of any specific point of view, but to demonstrate that visuality does not exist as such. My aim is to expose how visuality consists of an intricate intertwining of the one seeing and what is seen as a result of which we always see more, and always see less, than what is there to be seen. Moreover, that this one seeing is always necessarily a body. That, therefore, the question is not how to put the disembodied eye/I back into a body, but rather how is this body/subject that we are cultured to see according to the parameters of the Cartesian disembodied eye? How to conceive of 'just looking' as a necessarily impure and always synaesthetic event that takes place in a body as the locus of intertwining of various perceptual systems?

I approach the relationship between the body/subject seeing and what this body/subject sees starting from a selection of theatre performances that use *retheatricalization* as a strategy to expose the relation between what is seen and the bodies/subjects involved in seeing it.[8] These theatre performances use the means of the theatre to point attention to aspects of this relationship that usually remain invisible. Instead of using strategies of de-theatricalization to produce an illusion of the real, these performances use explicit theatricality to evoke reflection on what might be called the construction of the real. Doing so, they (like Brook's *The Man Who*) demonstrate how theatre as 'vision machine' can be used to manage attention in such a way as to evoke an inversion of the strategies described by Crary. The explicit theatricality of these theatre performances highlights aspects of various strategies of managing attention as they are part of ways of showing and telling. This theatricalization at the same time demonstrates their working *and* undermines their effect.

These theatre performances act as my *theoretical objects*.[9] I demonstrate how they can be read as theoretically meaningful statements embodied in the artistic discourse of the theatre. In this artistic discourse, thought 'moves' in different ways than in the theoretical discourse of the academy.[10] Each discourse has its own possibilities for showing and telling, for taking its audiences along, and for making these audiences move in response to the address presented to them. My aim is to make these differences productive.

I speak of 'theatre performances', thus combining two terms that for a long time have been (and sometimes still are) considered to be opposites, even antagonists. I choose to do so speaking from a theatre practice where this opposition is no longer productive. On the contrary, reiterating this opposition seems to get in the way of understanding how, in much theatre of the past decades, the influences of performance have been incorporated to a point where this has changed the whole notion of theatre. This development involved a moving away from drama as the central structuring principle of theatrical performance and towards a theatrical practice that combines elements of theatre and performance with dance, visual arts and music. Hans-Thies Lehmann describes this development beautifully when he refers to the transition from a *logocentric* way of structuring theatre performances by means of drama towards what he calls 'textual landscapes' (Lehmann, 1997, p. 59). In his later *Postdramatisches Theater* (1999),[11] he describes these same changes in terms of a 'multiplication of frames' that cancels out the operation of the single frame once provided by the dramatic structure. This causes ambiguous and confusing experiences, which Lehmann proposes to understand in terms of a political act. Political not because of what is represented on stage, but because of the ways in which the post-dramatic theatrical event draws attention to the problem of representation itself, to representational forms and of how they are perceived, or not.

Often, these theatre performances contain self-critical reflections on the semiotic habits that make up the discourse of the theatre. Usually, they are quite explicit about their own status as theatrical sign. In *The Man Who*, for example, no attempts are made to conceal the fact that the same actor embodies different patients or that the actor represents the patient rather than being one. Consequently, the performance as a whole seems to double the explanatory gestures of the doctor represented within the individual scenes. This doubling turns the explanatory gestures of the doctor into a set of revealing gestures which show the truth-speak of medical discourse to be a particular type of performance rather than objective observation of 'how things are'. At the same time, the reality of the actor (as distinguished from the character represented) is shown to be part of the discursive argument presented by the performance, rather than a 'real' or 'authentic' presence.

Staging the construction of the real

'Rather than define theatre as an unchanging identifiable object in the real, we might rethink it as a culturally conditioned mode of staging the construction of the real' writes Barbara Freedman in *Staging the Gaze* (Freedman, 1991, p. 50). The theatrical *apparatus*[12] as 'vision machine' stages ways of looking that respond to a particular culturally and historically specific spectator consciousness. Freedman writes about Shakespearean comedy and its relation to the Elizabethan world picture but her definition of theatre as a staging of the construction of the real seems to be valuable for rethinking the relationship between theatre and audience in other times and places as well. Freedman points to the relationship between theatre and the historical reality to which this theatre belongs but *without understanding theatre in terms of a representation of this reality*. Rather, theatre and reality appear as parallel constructions appealing to similar ways of looking. Theatre presents a *staging* of the construction that is also constitutive of the real. This staging responds to a similar spectator consciousness as implied by the construction of the real, while at the same time it is different; it is a theatrical staging. This ambiguous tension between similarity and difference brings Freedman to a definition of theatricality as 'that fractured reciprocity whereby beholder and beheld reverse positions in a way that renders a steady position of spectatorship impossible. Theatricality evokes an uncanny sense that the given to be seen has the power both to position us and displace us' (Freedman, 1991, p. 1).

With this definition of theatricality, Freedman points to the power of theatre to position us as viewers. The theatre addresses us and this address implies a position for us as subject of the vision presented to us. This does not mean that we as actual individual spectators necessarily share this vision or necessarily recognize ourselves in the position of subject of vision presented to us. This is precisely the ambiguous tension Freedman writes about: the address presented by theatrical performance has the power to both position us and displace us. The position of subject of vision as implied within the address presented to us mediates in our relationship to what is seen, and in this sense positions us, but as viewer we do not necessarily identify with this position. It can also cause a sense of displacement.

It is therefore important to distinguish between three different subjects involved in the event of visuality. In the first place, there is the subject seen. Second, there is the individual person as subject seeing. And third, there are the positions of subject of vision mediating between the two. In Chapter 2, I introduce the narratological term *focalization* to describe the relationship between these subjects of vision and what is seen. Focalization is a most useful concept for an analysis of the interaction between actual viewers and the visions presented to them in the theatre because it allows for an understanding of this interaction as a dynamic process of address and response in which the address presented by the theatre mediates in an event that for its actually 'taking place' depends just as much on the response of a particular viewer.

Focalization points to the relationship between ways of showing people, situations, and events, and the subjective point of view from where they are seen this way. Therefore, an analysis in terms of focalization can be used to expose how these ways of showing respond to viewers marked by particular presuppositions, experiences, fears, and desires; to viewers marked by ideas and presumptions characteristic of a particular world view. This may contribute to understanding why particular viewers would be willing to take up this point of view and recognize the vision presented as 'how it is', but may also explain why other viewers feel disoriented, alienated, or displaced by particular ways of showing. Or both at the same time, as in Freedman's definition of the ambiguity involved in theatricality. I will use focalization for precisely such an analysis of the relationship between the ways of showing and telling typical of both the dramatic theatre and its deconstruction, and a particular type of viewer. This viewer is the product of the history of Western Modernity and its various regimes of subjection, while being also post-modern in his or her fascination with the destabilization of this world picture and the modes of perception constitutive of it. My starting point will be the comparison of drama and perspective made by Hans-Thies Lehmann in *Postdramatisches Theater* (1999).

Drama, according to Lehmann, is a particular aesthetic logic at work in dramatic texts as well as in performances. Like perspective in painting, dramatic structure functions as a framework that presents the audience with a perspective on what is there to be seen as a result of which the audience knows how to look and how to understand what it sees. The dramatic perspective is *teleological*; it provides order in view of a goal or *telos* and corresponds to a world view characterized by unity and coherence in view of purpose and reason. In the post-dramatic theatre, this framework gets deconstructed or rejected altogether. Singular elements of theatre performance lose their connection with the totality that made the sensory meaningful. The deconstruction of the dramatic perspective, according to Lehmann, returns these singular elements to themselves as here and now and intensifies the presence of their sensory qualities.

With this comparison of drama and perspective, Lehmann presents an approach to both dramatic theatre and its deconstruction that does not focus primarily on text as the central element. Like Freedman, Lehmann understands theatre and reality as parallel constructions rather than as original and copy, or originary presence versus representation. The parallel observed by Lehmann, between dramatic structure and perspective in painting, points attention to the similarities between the two as strategies of focalization. His comparison brings to mind a history of producing vision and theorizing it: the history that is constitutive of the opposition of the modern disembodied subject versus an objectively visible world, as well as the separation of vision from the body and the other senses. Lehmann's comparison of drama and perspective allows for inscribing the typical strategies of focalization deployed in the dramatic theatre in this history. His comparison also presents a most useful analytical tool with respect to developments on the post-dramatic stage, pointing to the history that produced us, the spectators, as the kind of viewing subjects that are currently finding themselves challenged by the post-dramatic theatre. However, at this point Lehmann's account of the parallel between drama and perspective raises some questions as well. On the one hand, Lehmann's analysis of the deconstruction of dramatic structure bears witness to a Derridean critique of logocentrism. The deconstruction of the teleological perspective as given within the dramatic framework, Lehmann argues, does not result in direct access to the plenitude of being but instead evokes what he calls a multiplication of the frames mediating in how we see what we see. This multiplication undermines the effect of one single frame provided by the logocentric structure of dramatic structure and this way suspends the meaning-making function of dramatic frame. At the same time, however, does Lehmann's description of the effect of this multiplication in terms of 'increased perceptibility'[*gesteigerte Wahrnehmbarkeit*], intensified presence and essence of its sensory qualities [*gesteigertes Hier- und Sosein seiner sinnlichen Beschaffenheit*] and the 'singular led back to itself'[*das Einzelne auf sich zurückgefürt*] suggest that on the post-dramatic stage we finally are able to see what is there as it is in itself; that the source of the confusing experiences is located firmly in the thing as it is given over there. (Lehmann, 1999, p. 290). The result is that perception on the post-dramatic stage manifests itself in a paradox: the multiplication of frames results in the increased perceptibility of the thing in itself.

This paradoxical situation, of which Lehmann's account is exemplary, will be subject to a twofold approach. On the one hand, I will use Lehmann's comparison of drama and perspective as a starting point for an exposure of dramatic theatre and its deconstruction as culturally specific practices of managing attention responding to the desires, expectations, presuppositions and fears typical of the modern scopic regime. On the other hand, I will present a critical reading of how critical and theoretical responses to the deconstruction of the dramatic theatre show these to be expressive of these very same subjective desires, fears, presuppositions and expectations. I show that the deconstruction of the dramatic theatre as well as many of the theoretical responses to it, are themselves manifestations of what Hubert Damisch (1995) has called the 'perspective paradigm' rather than a move beyond perspectivalism.

Perspective as key to the locks of our senses

Perspective as invention of the early Modern period – first schematized by Alberti in 1435 – is fundamental to the development of the modern scientific world view and the constitution of the modern scopic regime. Far more than being just another technique of rendering images, perspective has considerably effected conceptions of the visible world, the relationship between this world and observers, as well as conceptions of what it means to see and how seeing is related (or not) to the other senses and to the body. Vision as presented by the perspectival image is based on what Norman Bryson has termed the differentiation of the corporeal glance from the idealized, disembodied and monocular gaze.[13] The viewer is offered a position from where the image produces an 'eternal moment of disclosed presence' as an effect. This presence-effect – and not correspondences between image and reality depicted – turns the image into a convincing representation of 'how it is'.

With his analysis, Bryson argues against the persistent notion of the perspectival painting as a copy of reality and linear perspective as a technique developed to capture a natural perspective, perceived in the world seen around us. More precisely, he uses his analysis to criticize art historian E. H. Gombrich's conception of a painting as a record of perception (*Art and Illusion*, 1960).[14] Gombrich's conception of painting as a record of perception, Bryson argues, is itself a product of the Renaissance treatises that inaugurated a tradition of presenting perspective as a way to represent the world according to laws of nature and perception. Today, rather than promising a *finestra aperta* on the world, the artificial perspective of the early Renaissance is clearly visible as a technique of producing images. Moreover, this is a technique that can be located in time and place, and that can be historicized. It appears to be much harder to see the notion of perspective itself as a historical invention, and to grasp the ways in which this invention has pervaded our conception of the visible world.

Mitchell likewise reminds us of Gombrich and his conception of pictorial illusionism as providing 'keys to the locks of our senses'. Mitchell argues that if vision itself is to be understood as a product of experience and acculturation, then what we are matching against pictorial representations cannot be any sort of naked reality. Instead it must be 'a world already clothed in our systems of representations' (Mitchell, 1986, p. 38). This means that perspective, rather than providing the key to the locks of our senses, must itself be understood as part and parcel of the way our senses are locked in. However, as I argue, this also means that perspective might prove to be useful as a key after all, namely in understanding how our senses are cultured to perceive certain privileged modes of representation as more natural, real, objective, or convincing than others, and to relate these effects to the discourses which mediate in what we think we see.

A brief look at the entry on perspective in the dictionary suffices to illustrate how the notion of perspective, far from being merely a technique of making images, is intricately intertwined with ideas, ideals, and ideology. The word 'perspective', according to the *OED*, is used to refer to the science of optics or the instruments used in performing this science, like a spyglass, a magnifying glass, or a telescope.[15] It may also apply to cases where 'the art of delineating solid objects upon a plane surface produces the same impression of apparent relative positions and magnitudes, or of distance, as do the actual objects when viewed from a particular point', or 'the appearance presented by visible objects, in regard to relative position, apparent distance etc.' (p. 606). Perspective can thus mean something perceived or discovered in the 'natural' world, as well as a technical means of making such a discovery. Perspective can be used not only to refer to the technique used to produce such a vision artificially, but also to refer to the vision presented by this technique.

Perspective is used to describe something actually seen, as well as 'the relation of proportion in which the parts of a subject are viewed by the mind; the aspect of a matter or object of thought, as perceived from a particular mental "point of view"' (p. 606). And finally, perspective can refer to either something 'natural', 'as it really is', or to 'a picture so contrived as seemingly to enlarge or extend the actual space, as in a stage scene, or to give the effect of distance', or even to describe 'a picture of figure constructed so as to produce some fantastic effect; e.g. appear distorted or confused except from one particular point of view, or presenting totally different aspects from different points of view' (p. 606).

Perspective presents an image of how these different uses are intertwined in the way the concept is used. 'Get things in perspective' is used as a metaphor to describe seeing things in their true relative proportion, which is actually an odd metaphor since getting the 'right' size, is precisely what perspective falsifies. In a perspectival drawing, this distortion is performed in order to give the *illusion* of these true proportions as they would appear from a particular point of view. Exactly for this reason, the concept perspective can also serve as a pointer, *conceptual metaphor* or *searchlight* (Bal, 1994, p. 40) that draws attention to the relationship between such illusions of objectively given world that exists as stable entity independent from any particular point of view and the subjective point of view from where the world can appear as such.

Concepts are not simply a neutral way of naming things since they influence the way the world gets constituted through them. In this sense, Bal argues, concepts are like metaphors replacing a story, and these stories replaced by them are told from a subjective point of view. Often, these metaphors are so deeply integrated in our thinking that they are not frequently recognized as such. They present us with basic cognitive categories that organize thought and structure how we act, think, and imagine. As a result, the story replaced by these metaphors (as well as the point of view implied by this story) is forgotten and the metaphors become constitutive of how the world appears to us. Perspective, I argue, is such a metaphor covering a story. Perspective involves much more than what meets the eye. Perspective is a symbolic form, expressive of, as well as productive of, a particular conception of the world. As such, its function is not specular or passive, but constitutive within the register of representation, of the order and meaning of things, and of the 'world of objects'. Perspective is informing perception rather than corresponding to it. For this reason, Hubert Damisch proposes to speak of perspective as a 'constellation of ideas, beliefs and prejudices that imposes their law in a given period on thought as it is expressed in art and science' (Damisch, 1995, p. 26). Perspective is integrated into our knowledge at the most implicit or unconscious level and it is symbolic precisely in the sense that the subject is absorbed in it and produced by it (Damisch, 1995, p. 19–20).[16] As a regulatory structure, it is sometimes in operation precisely where one least expects it, where its intervention is least visible.

Procedure

The institution of perspective theatricalizes the field of vision. It creates a scenographic space in which all that is seen is staged for a viewer. Paradoxically, despite the high degree of scenic manipulation required to successfully integrate the rules of perspective into a painted or otherwise constructed scene, the promise of perspective is that of immediacy. There is direct access to something that appears 'realistic', and, seemingly, no boundary between the space I stand in and the space it presents. Similarly, the aesthetic logic of the dramatic theatre presents the audience with a stable and detached point of view, allowing spectators to project themselves into the onstage world. This simultaneously brings spectators closer to the world onstage, while creating a distance from their bodies as the loci of their looking. As I demonstrate, the deconstruction of dramatic theatre can be seen to serve a similar purpose, evoking as it did the effects of immediacy and immediateness just at the same moment that dramatic theatre lost its power to absorb an audience into its onstage world. Like the perspectival construction of dramatic theatre, strategies of deconstruction aim at bringing the spectator closer to what is seen on stage, breaking up and taking away the dramatic frame that mediates between the one seeing and what is seen. The spectator is no longer presented with a fixed perspective, and is free (at least metaphorically) to wander. I argue, however, that this wandering around again involves the spectator, in a position as the subject of vision, at a certain distance from him or herself as a body looking.

In Chapters 2–4, I trace this paradoxical nature of perspective – this paradox that is perspective – as it is part of the visuality that takes place both within the dramatic theatre and within the dramtic theatre's deconstruction. In Chapter 2, the analogy between deixis and perspective is the starting point for an analysis of the address presented by theatrical performance in terms of positioning. I introduce the concept of 'focalization' to describe the ways in which such positions present the audience with points of view that mediate within the vision presented on stage. I demonstrate how the analogy between deixis and perspective allows for an understanding of both the dramatic theatre of the past and its deconstruction on the 20th-century stage in terms of aesthetic strategies aiming at what Michael Fried (after Diderot) terms *absorption*. Fried's work also illuminates the relationship between such anti-theatrical tendencies and what might be called metaphysical thinking in the field of vision. In Chapter 3, I take a closer look at how point of view as implied within the address presented by dramatic and post-dramatic theatre mediates in visions of contemporary and historical reality presented on stage. I demonstrate how Lehmann's comparison of drama with perspective allows for an understanding of the power of persuasion of theatre performance as the result of the invisible logic that shows us what is seen according to a particular point of view, while at the same time this point of view is precisely what goes unnoticed. In Chapter 4, I argue against the mistaken idea that deconstructing or rejecting dramatic frame leaves spectator free to see as he or she wishes.

In these chapters, the focus is on exposing the relationship between what can appear as 'just looking' and the aesthetic logic of the theatrical event. I argue that perspective (and not the absence of it) as a definition of the complex relationship between a spectator and the world seen 'as it is', presents an alternative to the binary opposition of representation and presence dominating much of the discourse surrounding contemporary theatre performance. This alternative exposes the persistence with which this opposition is reiterated as 'a nostalgia for the present' (Freedman, 1991) and as the effect of the persistent denial precisely of the self, the body seeing, as the locus of looking.

The chapters that follow present a further exploration of this relationship between what is seen in 'just looking' and the body involved in seeing things 'as they are'. I engage more fully with the close relationship between the theatrical apparatus (the paradigmatic example within modern practice for managing attention), representational thinking (including the opposition of representation versus presence) and the psychoanalytical notion of 'the subject'. A recurring motif in these chapters is the Lacanian model of mirror stage identification, and my argument for the usefulness of Lacan's model for the analysis of visuality as it occurs in the theatre. Lacan's analysis, if read as a culturally and historically specific explanation of how we relate to what we see, illuminates the relationships between the scopic regime of modernity, the psychoanalytical subject, the disembodied eye/I, and the theatre as a practice of staging vision.

Chapter 5 elaborates parallels between what Barthes terms the *punctum* – in photography – and Freud's notions of the *slip of the tongue* and the *navel of the dream*. The *punctum*, according to Barthes, is a detail that sticks out from the frame of symbolic reality. I argue that such a detail can also be read as a meaningful symptom, drawing attention to the relationship between seer and seen in what seems to be 'just looking' at what is 'there to be seen' onstage. In Chapter 6, I use this approach for a closer look at the relationship between the perspective paradigm and the gender-asymmetry typical of psychoanalytical accounts of subjectivity, and how this asymmetry in psychoanalytic practice obscures an invisible logic at work in accounts of the world 'as it is'. Here, perspective as conceptual metaphor is used to point attention towards the blind spots that are constitutive of the 'just looking' of the disembodied eye/I. In Chapter 7, I introduce Silverman's reading of Lacan to address the idea of seeing as an action mediated by a visual unconscious: the cultural repertoire of images. Silverman here presents an alternative to the way film theory has elaborated the gaze. Silverman's argument is useful in analysing visuality in the theatre, while also allowing for a historization of the Lacanian mirror stage model. I demonstrate this using a staging of Heinrich von Kleist's text *On the Puppet Theatre* to illustrate my reading of Lacan's text on the mirror stage. In Chapter 8, I turn to the historical Anatomy Theatre for a critical reading of the relationships between the body image provided by anatomy, the Cartesian disembodied eye/I, and the subject of the mirror stage. In Chapter 9, I point to the connections between the mirror stage and what Crary describes as the history of managing attention in modernity. Read as a response to disembodied vision, Lacan's mirror stage essay illuminates how psychoanalysis naturalizes a historically specific conceptualization of perception. This historicization turns the theatre into a staged version of the ways in which, as Lacan puts it, the modern subject is caught, manoeuvred and captured in the field of vision (Damisch, 1995). This brings me to a reconsideration of what we think we see (as a product of being bodies involved with the world through various simultaneously functioning perceptual systems), and to seeing as an activity that takes place at the intersection of the physical possibilities of bodies, and how these are shaped by cultural conditioning.

In Chapter 10, the final chapter, I propose a mode of approaching analysis of visuality in the theatre starting from the ways in which the theatre addresses the seer; how the cognitive-perceptual perspectives presented to this seer suggests particular positions in relation to what is seen, positions that may resonate, interfere, or contradict each other; how these positions may invite stories, associations, interpretations, in which self and other get mixed up or conflate. I demonstrate how various concepts introduced so far, may support such analysis of visuality as it takes place in the theatre. I do so by means of an analysis of several scenes from four different versions of one and the same opera, which is Wolfgang Amadeus Mozart's *Die Entführung aus dem Serail* (*The Abduction from the Seraglio*), an opera that stages seeing otherness and has a history of being criticized for the way it does so.

The seer

To denote the one doing the seeing in the theatre I will from now on use the term 'seer'. A more obvious choice would have been to speak of the 'spectator'. The disadvantage of the word 'spectator', however, is that it has come to be associated with passivity, the onlooker at a spectacle, gaping at the given to be seen: reason for Crary to propose 'observer' instead, a word with which he wants to stress how the one who is seeing is always doing so from within a prescribed set of possibilities, one who is embedded in a system of conventions and limitations (Crary, 1992, pp. 5–6). Crary's approach proves to be most useful in describing how this 'observer' and his or her ways of seeing are the product of cultural practices that condition how this person will see. This helps to understand why learning to see is, as Peggy Phelan terms it, 'training careful blindness' (Phelan, 1993, p. 13). We always see less than is there.

But with this term 'seer', I want to point to the opposite as well. We also always see more than is there. The term 'seer' apart from meaning 'the one who sees' and 'an overseer, an inspector', is also associated with insight, revelations, prophecy, second sight, and magic.[17] The seer is someone who sees things that are not there: future things, absent things. Seeing always involves projections, fantasies, desires and fears, and might be closer to hallucinating than we think.

With the term seer, I acknowledge that there is no way of opening our eyes to what is actually there to be seen. We are always 'seeing things'. But, although we are much less free in what we see than we may think, we are also much freer than we think, because the subjectivity of vision opens up the possibility of change and transformation. The term 'seer' is an acknowledgement of the fact that we always see more or less than what is there and that, therefore, seeing is always affected by with ideals, values, presuppositions, fears, and desires. These factors do not necessarily match our own, nor those of the ones we see. But the term 'seer' is also an acknowledgement of the possibility of opening our eyes to difference.[18]

2
Step Inside!

Perspective, Damisch argues, is neither a language nor a code. Nevertheless, perspective has some things in common with language and especially with the phenomenon linguistics has termed *deixis*. In language, indicators of deixis are the personal pronouns ('I' and 'you') as well as the demonstrators, adverbs and adjectives that organize the spatial and temporal relationships around the subject. This 'I' and 'you', 'here' and 'there', are empty forms which each speaker, in the exercise of discourse, appropriates to him- or herself and with which he or she relates to his or her person. Similarly, Damisch argues, perspective 'institutes and constitutes itself under the auspices of a point, a factor analogous to the "subject" or "person" in language, always posited in relation to a "here" or "there", accruing all the possibilities for movement from one position to another this entails' (Damisch, 1995, p. 53).

In theories of the theatre, deixis has proved to be most useful for the analysis of dramatic speech, and the ways in which this speech establishes character, space and action. Deixis describes the way relationships are set up between persons and other persons, persons and objects, here and there, earlier and later. An analysis of the function of deictic markers in dramatic or performance text therefore illuminates how the world onstage 'spaces-out' as a result of what is called the system of internal theatrical communication, that is the communication between the characters in the fictive cosmos represented onstage. In this chapter, I argue as well for the utility of deixis when analysing the system of external theatrical communication. I demonstrate how Damisch's analogy between deixis and perspective illuminates the interaction between the world onstage and the seer. I discuss how such an analysis does not imply reducing the visual to the verbal, but rather, helps clear up the confusion concerning similarities and differences between the verbal and the visual. Further, I point out how this may illuminate the ways in which the verbal and the visual interact and are intertwined. The analogy of deixis and perspective thus presents a first step towards a model that can account for interaction between seer and seen in the theatre in terms of the interaction between the multimedia address presented by the theatre and a seer involved with the world through the various perceptual systems.

Second, such an analysis of how visuality in the theatre 'takes place' (as the result of the interaction between the address presented by the aesthetic logic of the theatre and the response of a seer) allows for the inscription of modes of looking as they are taking place in the theatre within the history of visuality as a culturally specific phenomenon. Here, the analogy between deixis and perspective presents a point of connection between modes of looking invited by dramatic and post-dramatic theatre and a history of de-theatricalizing vision in painting. Bryson (1983) observes how in painting the suppression of signs of deixis supports the illusion of a moment of disclosed presence, of a vision of a world that exists as stable entity independent from our point of view. Similarly, suppression of signs of deixis in the external system of theatrical communication supports the illusion of a world on stage that exists independently from our position as audience, a world we are looking into, rather than implicated in. A world experienced as if through Alberti's *finestra aperta*: a window opened by the proscenium arch.

Such an audience position has been criticized by twentieth-century theatre makers aiming for more direct and intense contact in the theatre between seers and seen. Many have argued for (and practiced) a deconstruction, transformation or rejection of the conventions of the dramatic theatre of the past, sometimes rejecting the theatre altogether in favour of what was perceived as the more immediate presence of performance. In this discourse, as Elin Diamond (1996) observes, 'theatre was charged with obeisance to the playwright's authority, with actors disciplined to the referential task of representing fictional entities' while spectators are similarly disciplined and 'duped into identifying with the psychological problems of individual egos and ensnared in a unique temporal-spatial world whose suspense, reversals and deferrals they can more or less comfortably decode' (Diamond, 1996, p. 3). Performance, on the other hand, has been honoured with 'dismantling textual authority, illusionism, and the canonical actor in favour of the polymorphous body of the performer. Refusing the conventions of role-playing, the performer presents her/himself as a sexual, permeable, tactile body, scourging audience narrativity along with the barrier between stage and spectator' (Diamond, 1996, p. 3).

Diamond's description is ironic, indicating that, today, we tend to take a certain distance from the celebration of unmediated presence and directness that characterized early performance theory. The term presence, when used at all, is placed between quotation marks or replaced by notions like 'presence effect'. 'Presence' is now understood as necessarily rhetorical and always relying on representation. That is, presence relies on other signifiers and thus remains within the realm of the already constructed. Derrida's 'always already' has left deep marks. Yet, it still appears to be difficult to avoid conceptual oppositions like representation and presence, meaning and materiality (and the list goes on) when confronted with the strategies used on the contemporary stage. Not the least because these theatrical and performative strategies seem to get right to the heart of such oppositions, playing with them and diametrically playing out one term against the other. Moreover although the notion of an ontological foundation for these oppositions may now be highly problematic, as effects they are most present and most impressive.

At this point, the analogy between deixis and perspective presents an alternative approach that allows for an understanding of both the dramatic theatre of the past and its deconstruction on the twentieth-century stage in terms of aesthetic strategies aiming at what Michael Fried (after Diderot) terms *absorption*. In his *Absorption and Theatricality: Painting and Beholder in the Age of Diderot* (1980), Fried discusses the relation between painting and beholder as it develops during the middle and late eighteenth century in France. During this period, a certain tension made itself felt between, on the one hand, the painting as representation (i.e., as an object produced to represent something for someone) and on the other hand, a certain uneasiness with this condition of representation. Fried refers to the moment this paradox appeared as 'a momentous event, one of the first in a series of losses that constitute the ontological basis of modern art' (Fried, 1980, p. 61). This loss inaugurated a continuous search for representational strategies that suggest the absence of a beholder. According to Fried, the seminal figure for understanding the beginnings of this tradition is Denis Diderot, whose writings on drama and painting have at their core the demand for the achievement of a new and paradoxical relationship between the work of art and its audience.[1] His texts on theatre, his Salons and his related texts on painting all deal with the conditions that have to be fulfilled in order for the work of art (painting or theatre) successfully to persuade the audience of what he (Diderot) calls the truthfulness of its representation. In order to appear as truthful, it is crucial that the beholder be treated as if he or she is not there. Like perspective, such strategies, aiming at absorbing the seer into the seen, need to go unnoticed in order to achieve their effect. As soon as they become recognizable, the effect turns into its opposite, which is theatricality.

Absorption does not refer to an a-historical quality of a work but is the effect of the interaction between a work of art produced at particular time and place, and a historically and culturally specific viewer. The eighteenth-century paintings that Fried discusses may have rather different effects on twenty-first-century viewers, or on viewers from a cultural background that differs in important respects from that of the French upper class for whom they were originally intended. Fried, therefore, refers to the accounts of eighteenth-century art critics to point out how and why these paintings, at the moment of their production, were conceived of as absorptive. He also uses their critical responses to demonstrate how strategies aiming at absorption eventually lost their power to achieve absorptive effects, were denounced as *theatrical* and had to be replaced by new strategies, not immediately visible as such. In addition, Fried presents a perspective on anti-theatrical tendencies in late-modern art, as well as the discourse surrounding it, including his own earlier work 'Art and Objecthood', a text that had played an important part in the discourse on the opposition of theatre and performance (the kind of discourse Diamond refers to).

In 'Art and Objecthood', Fried uses the notion of theatricality to criticize minimalism (or, as he calls it, 'literalist art') and to set up an opposition between the experience of minimalist works and that of modern art works characterized by instantaneousness and what he calls 'presentness'. 'Presentness is grace', is the final line of his text, for presentness lifts us above the perverted theatrical mode of being to which we are confined for most of our lives (Fried, 1968, p. 147). This 'theatrical mode of being' refers to the awareness of oneself as a spectator, a condition which Fried clearly dislikes. Therefore, he argues against works of art that explicitly address the viewer, dismissing them as 'theatrical'. Theatricality, in Fried's terms, produces the seer as subject in a way comparable to the way deixis functions to set up relationships in language, through an address that provides the seer with a position in relation to what is seen. Presentness, on the other hand, indicates the absence of such a relationship.[2]

Josette Féral (1982) and Chantal Pontbriand (1982), in two articles appearing in the same issue of *Modern Drama*, adopt Fried's terminology to explain the intensities of 'presentness' experienced in performance as a result of the deconstruction of dramatic representation, mediation and point of view characteristic of the theatre. Foreshadowing Lehmann's concept of the post-dramatic theatre as a non-teleological architecture,

Féral defines performance as a 'primary process lacking teleology' (Feral, 1982, p. 177). Performance involves a process of undoing predetermined points of view rather than constructing them. Since it tells of nothing and imitates no one, performance escapes illusion and representation to reveal what remains hidden behind the symbolic mediation that takes place in the theatre. 'Performance rejects all illusion', writes Féral (1982, p. 171). It 'presents, it does not represent' writes Pontbriand (1982, p. 155).

Inspired by Derrida's analysis of the opposition of speech as originating self-confirming palpability versus writing as a secondary, posterior activity, Féral, Pontbriand (and others with them) reject the illusory character of dramatic representation in favour of performance as a 'primary process' 'revealing what remains hidden behind symbolic mediation' (Féral, 1982) or presenting instead of representing, thus 'showing the real without mystification' (Pontbriand, 1982). This brings them to an understanding of performance in ways that reiterate precisely those binary oppositions that Derrida sets out to deconstruct. In their texts, as Auslander observes, Féral and Pontbriand use Derrida to further the modernist urge for reduction to an essence; in the search for timeless and immediate presence they dress up Greenbergian modernist aesthetics in poststructuralist clothing (Auslander, 1997, p. 56). At this point, Fried's later work allows for an alternative approach that illuminates the relationship between the aesthetic strategies deployed in both dramatic theatre *and* performance, and what might be called traces of metaphysical thinking in the field of vision, as well as with the relationship between such traces of metaphysics in the field of vision and certain modernist traditions in art.[3] As I demonstrate, this opens the possibility of an understanding of the presentness of performance not in terms of the absence or undoing of points of view, but as the effect of their remaining invisible. Furthermore, Fried's reading of Diderot suggests that the teleological architecture of dramatic theatre actually may be used to a similar end, namely of absorbing their audiences into an experience that erases the awareness of the relationship between what is seen and who is seeing, doing so by – intentionally or unintentionally – presenting the seer with an address that responds to culturally and historically specific desires, expectations, presuppositions and anxieties.[4] In the following, I demonstrate what such a reading might entail, in dialogue with *Artifact* (1984) by choreographer William Forsythe, a performance that reads as a preposterous staging of Lehmann's account of the development from dramatic to postdramatic theatre.[5]

Do you see what I mean?

Artifact is a ballet constructed as a late-twentieth-century commentary on the nineteenth-century story-ballet. The ballet is constructed around words. The words are arranged in a diagram and printed in the programme book (Figure 1). The diagram presents a key to the construction of the performance while also used within the performance as a tool for deconstructing language; the language of words as well as the multimedia language of the stage. In *Artifact*, this language of the stage is what is at stake. Like most theatre performances, *Artifact* invites the audience to make sense of what is presented to them. *Artifact* self-consciously presents itself as an artefact (as opposed to the Kantian self-contained, autonomous work of art) that has been constructed to present its address to an audience. This address is neatly summarized in the question: 'Do you see what I mean?' directed towards the audience by one of the characters onstage.

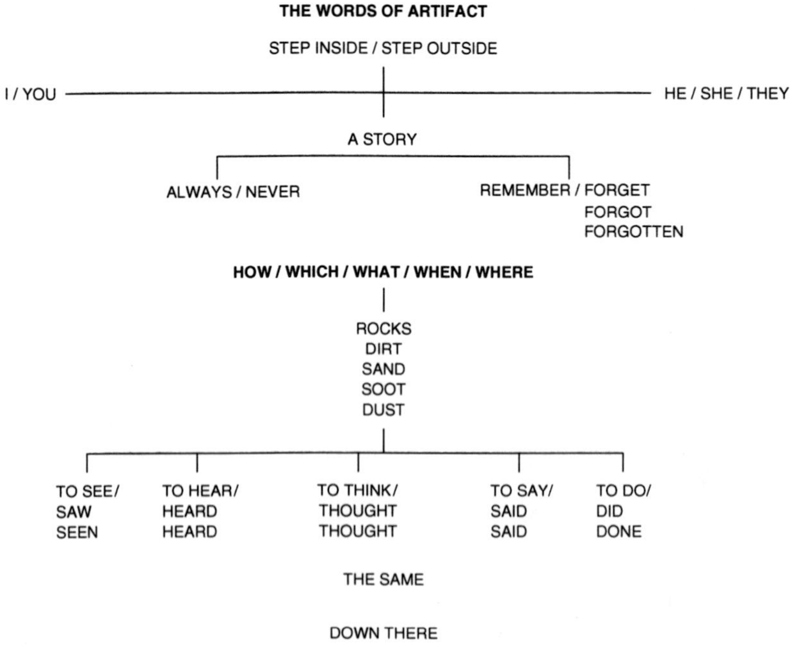

Figure 1 The Words of Artifact by William Forsythe. Reproduced with his permission.

With this witty question, *Artifact* highlights the relationship between stage and auditorium, a relationship that often remains implicit or is even explicitly denied in more conventional forms of theatre. The question 'Do you see what I mean?' calls this relationship into question and presents a critical commentary on developments in theoretical approaches to theatrical meaning production occurring around the same time *Artifact* was created. These developments amounted to a shift in emphasis from a structuralist' approach towards dramatic or performance texts to the pragmatics of theatrical communication. This shift extended the Sausurrean model of signifier/signified by drawing attention to the *interpretant*, the necessarily third element in signification. The *interpretant* was given more attention by Peirce. Reception is built directly in into Peirce's famous definition of the sign or *representamen* as 'something that stands to somebody for something in some respect or capacity', the *interpretant* being 'the equivalent of that sign created in the mind of that person' (Peirce quoted in Nöth, 1990, p. 42). Since every sign creates an *interpretant* that in its turn is the *representamen* of a second sign, semiosis results in an infinite series of successive *interpretants*. There is no first and no last sign, but rather a continuous process that can be interrupted, but never ended. This is a process, furthermore, in which what *you* see does not necessarily correspond to what *I* mean.

But *Artifact* has more to offer than a clever illustration of the fact that a sign can only stand for something in relation to somebody to whom it appears as a sign. In *Artifact*, the relationship between this *I* meaning and *you* seeing is at stake. *Artifact* reveals a demonstration of how the 'I' and the 'you' implied in this address – as well as the relationship between them – are constituted in and through the address. *Artifact* therefore presents a critique of semiotic models based on a static model of communication, in which a series of coded messages are sent or enacted, so that they may be subsequently received and decoded by a spectator. The turning away from such models of theatrical communication resulted in a variety of approaches, usually gathered under the rubric of pragmatics. In this context, the most widely and frequently invoked analytical framework (the true 'Prince of Pragma' as per Keir Elam, 1988), is speech-act theory as developed by J. L. Austin and John Searle.[6]

Austin's observation that, in many cases, saying things is doing things with words rather than using these words to refer to some absent state of affairs, confirms what theatre makers have known for some time. Moreover, doing things with words is central to the functioning of dramatic text. It is little wonder then that speech-act theory proved productive in analysing how meaning results from how things are done with words onstage. More complexly it appears to be bringing Austin's speech-act theory to bear on the external system of theatrical communication, that is, the transactions between stage and audience. This is the level of communication that the woman in *Artifact* draws attention to with her question 'Do you see what I mean?' Elam refers to Ross Chambers' (1980) proposal to understand theatrical performance as an overall communicative act (analogous to the speech act) that encodes an offer or invitation to the audience, an invitation bearing the message 'come and interpret me' (Elam, 1988, p. 46). According to Elam, however, this approach proves unproductive. First, this understanding of the address presented by the performance converts the overall act created by the performance into a tautological exercise: the invitation is the founding convention of the theatrical transaction itself. Second, understanding a performance as an address presented to a spectator to 'come and interpret me' runs the risk of again reducing the spectator to a decoding machine, and thus denies the work involved in responding to the address. He refers to Barthes' (1993) distinction between the *studium* (as 'a sort of alert but not particularly intense being interested') and the *punctum* (as 'an irresistible pricking or injury') to conclude that much of what goes on under the name of theatrical pragmatics (and theatre semiotics in general) is limited to the zone of the *studium* and keeps a safe distance from the tropics or dangers of the *punctum* (Elam, 1988, pp. 48–49).

I come back to Barthes' distinction and its possibilities for theatrical analysis in Chapter 6, where the *punctum* is part of my argument for the usefulness of semiotic analysis in analysing theatre, provided that semiotics shift the attention from what is 'to be seen over there', towards a relational model that takes into account the relationship between what is seen and the point of view from which it is seen. I show how *punctum*, in Barthes' terms, can serve as a meaningful instrument in exposing the relationship between what is seen and the subjective point of view from where it is seen. In this present chapter, I take a first step towards this argument by taking a closer look at what Elam calls 'the founding convention of the theatrical transaction', in other words, the spectator being called upon to interpret what he or she sees and hears. This (implicit) address to the audience is verbalized and made explicit by *Artifact*, yet reformulated in a crucial way.

Speech acts in the field of vision

'Do you see what I mean?' draws attention to two different activities involved in the act of 'come and interpret me'. This question addresses the audience as both listener and seer. Elam recognizes these two activities when he describes the act performed by the audience as being 'called upon to interpret what he hears and sees' (Elam, 1988, p. 47). In his summary of 'come and interpret me', however, Elam conflates these two into the single act of interpreting, and this obscures the way seeing and hearing imply different subject positions for an audience. *Artifact*, by re-separating the two, draws attention to an easily overlooked complication of the use of speech-act theory for the analysis of multimedia texts: that the different perceptual and interpretative activities involved in the response to such a multimedia address implies the existence of different, simultaneous, subject positions for the audience.

On the one hand, the question 'Do you see what I mean?' can be read as an invitation presented by the woman to imagine seeing things the way she does and, in this way, to understand what she means. In this instance, the audience is addressed as a 'you' hearing the 'I' speaking, but also invited to take up the position of the 'I' seeing onstage, seeing as if from her point of view. But 'Do you see what I mean?' can also mean 'Do you see what I (as a visual sign) mean?' In this case, the address presented to the audience is again an address to the seer as a 'you' being spoken to, but this 'you' is also addressed as the 'I' that sees her, the 'I' that is the subject of vision of the perspective that the performance constructs upon her. This conflation of the various positions involved in responding to the address presented to the audience is precisely what tends to get obscured by the ambiguous relationship between seer and seen as given in perspective. Perspective addresses the viewer with the invitation to take up a position from which to see. This position, however, is not that of a 'you' addressed but that of an 'I' seeing. This 'I' seeing is a position implied within the address presented to a seer. This position mediates in the relation between seer and seen; it is part of how their relationship is established. This process of mediation is called focalization, in narratological terms.

Internal focalization

Whenever events are presented, they are always presented from within a certain 'vision'. A point of view is chosen, whether what are being considered are 'real' historical facts or fictitious events. The narratological concept of focalization describes the relationship between this vision and that which is 'seen'. The concept of focalization is therefore comparable in many ways to that of perspective. But there are important differences. Perspective tends to focus attention only on what is seen, and to direct attention away from the position from which things are seen. In this way, the relationship between the thing seen and the determinism inherent in from where it is seen is obscured. Focalization, on the other hand, describes the precise relationship between the subject viewing and the object viewed, as it is given within the particular construction of the visual, verbal or multimedia text.[7]

Focalization draws attention to the position from which things, people and events are seen and also how this subjective position mediates the vision presented to us. Focalization helps to clarify how such subjective positions implied within the address presented to us by, for example, a theatre performance, invite us to take up these positions, identifying with the point of view they present us with. 'Identification' here does not mean the kind of non-critical, passive reception whereby the spectator imagines him- or herself to be a character represented on stage. The concept of focalization involves a type of identification that does not aim at erasing difference between seer and seen, but is more like what Bruce Wilshire (1982) metaphorically describes as 'standing in'.[8] The spectator is invited to 'step inside' and to take up a position as represented within the work and to see as if from there.

In *Artifact* several onstage figures act as internal focalizors, directing the attention of the audience by means of visual signs (like arms and looks pointing in specific directions). Focalization also happens through what might be called bare bones dramatic construction. At the beginning of the performance, the woman, in a Louis XIV-like period dress, comes forward, claps her hands and the music starts to play. Slowly, she approaches the audience, throwing kisses, making theatrical gestures of invitation and honouring the audience with elegant bows. Her dress brings to mind the Baroque theatre organized in relation to the vision of the king and her salutation is certainly worthy of a king. She performs the address called for by the architecture of a Baroque theatre. Having reached the front of the stage, she halts, stretches up, looks at the audience and says 'Step Inside' (Figure 2). She thus literally becomes an internal focalizor, explicitly inviting the audience to do what many performances implicitly assume will occur: abandoning the observer's position in the auditorium and imaginatively projecting oneself into the world onstage. However, the explicitness of her invitation makes it an ambiguous gesture that threatens to undermine what it pretends to install. Because, successfully projecting oneself into and becoming absorbed by the onstage world depends on the seeming absence of just that relationship that is highlighted by her performance. This relationship has to go unnoticed in order to present the spectator with the illusion of another world on stage, one that can be seen through Alberti's *finestra aperta*.

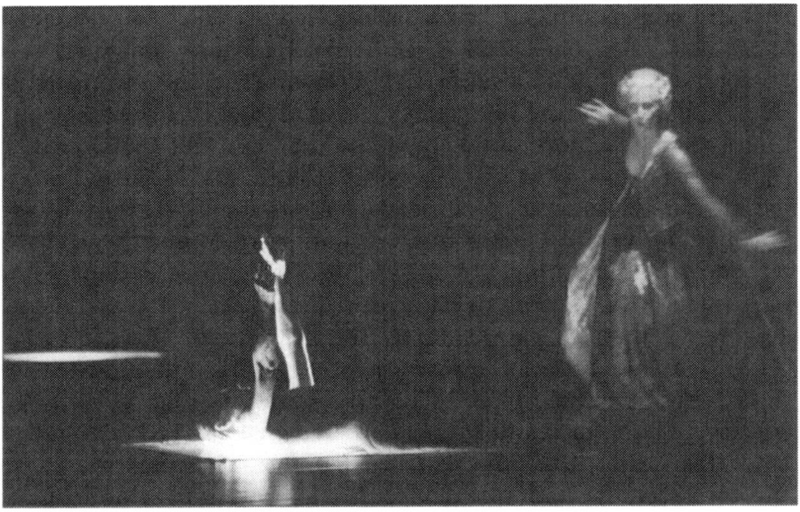

Figure 2 Photograph by Dieter Schwer of Kate Strong in Willam Forsythe's *Artifact*. Reproduced with permission of the photographer (www.Dieterschwer.com).

The woman highlights the theatrical nature of the situation with her invitation. She reaffirms that this is theatre by performing the critical gesture of Brechtian theatre: closing the gap separating stage and auditorium in order to create distance. Furthermore, something seems to be at stake in the order of things. In the case of *Artifact*, it is not the onstage landscape that seduces the audience to 'step inside'. Rather it is the invitation itself that marks the beginning of the action. This is the beginning of the theatrical event, seeming to demonstrate that it is only by the audience members assuming the point of view that the woman proposes that the event takes place, as though the positioning of the audience is part and parcel of the performance taking place. With her gesture, the woman frames what is there to be seen onstage as theatre. At the same time this gesture of framing consists literally of setting up a relationship between the seer 'over here' and the seen 'over there', inviting this seer over there to 'step inside', to leave behind reality and enter the fictive cosmos.

The woman in *Artifact* acts as the focalizor made of flesh, literally inviting the audience to see what happens from her point of view, to step into her shoes and see it as she does. *Artifact* also demonstrates how points of view implied within the address presented by a performance mediate in the unfolding of dramatic narrative. At first, the woman seems to be self-confident, even self-satisfied. Her text, all statements and orders, gives her an air of authority. Her voice, amplified by a microphone, is loud and clear and surrounds the audience from all sides. After a while, she is joined by a man dressed in a late-twentieth-century suit. The relationship between the man and the woman is constructed following various binary oppositions: man/woman, history/present, extravert/introvert, declarative/interrogative, and distance/closeness. The man, being a true antagonist, sets out to undermine the woman's position as wisdom broker. While the woman remains stationary, the man walks, asking questions like 'Which?' and 'When?' and 'Where?' and 'How?' through a megaphone. The megaphone, like the microphone, amplifies his voice, but instead of suggesting omnipresence, the source of the sound is firmly located in relationship to those he addresses. Furthermore, the megaphone amplification does not obscure, but rather emphasizes the distance between the speaker and his audience. Coming from different angles, his vocalization undermines the totalizing effect of her voice.

The man and the woman are involved in what seems to be an ongoing series of attempts to find out and describe what is the case onstage, and to convey this to the audience. In their attempts to do so they make use of the 'words of *Artifact*'. In the programme, these words are accompanied by dictionary definitions, as if to ensure their meaning. In the performance, however, reference becomes something of a desperate plea, a demand to the sign to convey Presence, Being, Meaning – which it simply will not deliver. They use the words of *Artifact* to produce grammatically correct sentences, using these sentences in ways that suggest that the phrases are meant to make sense of what is happening on stage. The degree to which the words relate to what is seen on the stage is however often difficult to grasp. The characters explore various possible combinations of words, as given in the diagram, using them again and again in different order, or repeating the same syntactical structure using different words. While reference becomes increasingly problematic, discourse on stage becomes deictic to the extreme. The performance thus seems to prove Benveniste's (1971) point that deixis, and not reference, is the essence of language. Although it is often hard to say what the words on stage refer to, the use of these linguistic signs does make sense as an address, an address that invites a response even though it is not clear what is meant. Indeed, when the woman asks the audience: 'Do you see what I mean?' her question is usually met with laughter.

While the question may be hilarious, it is also strikingly to the point. Audience members know what the words mean, but they have neither got the slightest idea what she means by them, nor what she, as a visual sign, is supposed to refer to. Nevertheless, her question is meaningful in that the audience is taken through a series of subjective transactions. Meaning and subjectivity come across through the play between 'me' and 'you' as positions produced by a function of linguistic signs, by means of an address through visual signs, through the directing of hands and eyes, through the choreography of bodies in space and through the constructions of perspective and point of view. The characters provide different perspectives on what is at stake and, since reference is highly problematic, it is not possible to say who is right and who is wrong. All attention is directed towards the mechanism of the dramatic conflict itself, revealing it to be a drama of vision and positioning.

External focalization

Apart from one or more internal focalizors there is always an *external focalizor*. This is the anonymous agent through whose eyes we as audience see the performance. As long as this position remains invisible, the performance can appear simply 'there to be seen' independent from any particular point of view. *Artifact* demonstrates how this 'invisible' position nevertheless mediates in the relation between seer and seen, and also how this mediator can be brought to awareness by de-naturalizing the position involved. To explain this, I will return once more to the invitation 'Step inside!' offered at the beginning of the piece.

With the invitation to 'Step inside' the woman reveals the performance to be an act of what Althusser calls *interpellation*. Interpellation is the speech-act of the social environment embodied in the state's representative, the policeman, who calls out 'Hey you!' causing the subject to turn around. This in turn constitutes him or her as subject. The policeman saying 'you' makes 'you' specifically into 'me,' that is, it makes me turn around, feeling addressed at the same time as I feel unsettled, taken out of myself, already in prison (Bal, 1999, p. 87).[9] In *Artifact*, it is the audience that is compelled out of its seat and into the world on the stage, and it is through this act of interpellation that the audience is constituted as the subject of vision.

During the second act, the safety curtain comes down with a bang, throwing the audience – having 'stepped inside' – abruptly back into their seats in the auditorium. With this powerful gesture, *Artifact* draws attention to the ambiguity at work in the original invitation. 'Stepping inside' describes an embodied action. While this phrase is used in this context to invite the audience to take a certain distance from their auditorium-located bodies and get closer to the stage world. This happens by means of an address to the seer in which this seer, bodily attached to his or her viewpoint, is at the same time released from this bodily 'locus of looking', as a result of which the relationship between what is seen and who is seeing gets obscured. The coming down of the safety curtain highlights precisely this relationship by disturbing the unproblematic identification with the point of view of the disembodied eye. As a result, instead of being released from his or her 'locus of looking', the seer is made aware anew of how he or she is bodily attached to it.

When, after only a few moments, the safety curtain goes up again, the scene on stage is radically altered. Just before the curtain comes down, a group of dancers move simultaneously in a symmetrical composition. When the curtain goes up, the dancers are still moving as if nothing had happened. Their formation in space, however, is completely different. This device is repeated several times. The new spatial composition of dancers no longer respects the conventions of visibility and frontality typical of ballet. Sometimes the dancers face the back wall or the wings. Sometimes compositions are partially or wholly invisible, taking place in the wings. At some point the dancers lie on their backs and perform their movements towards the ceiling.

Absorption

Through this remarkable gesture *Artifact* exposes the perspective at work in the vision presented. Suddenly, we become aware of the perspective implied within what until then seemed just there to be seen, and how this perspective provides us with a vision in relation to our point of view. Moreover, that this perspective needs to remain unnoticed in order for the seen to appear as just there, independent from our point of view. The piece shows how ways of staging support suggestions of visibility and accessibility, and also how, as a result, the attention of the audience is directed in such a way that he or she remains unaware of the specificity of this point of view, and unaware of him- or herself in relation to what is seen. Unawareness occurs through, for example, respecting conventions and responding to expectations concerning composition and frontality. *Artifact* also demonstrates how focalization can be used in such a way as to make the audience aware of this relationship and even to expose (aspects of) what is involved in this relationship: the expectations, desires and presuppositions that guide the looking of the seer. The effect is that the seer is no longer absorbed in the world presented on stage but made aware of this world as a staging, as theatre.

This relationship between the address presented to a seer, absorption and the theatre is the subject of Fried's *Absorption and Theatricality. Painting and Beholder in the Age of Diderot.* In this study, Fried distinguishes between three successive strategies aimed at achieving absorption in eighteenth-century French painting. In the early- and mid-1750s this effect was achieved, for example, by depicting people sleeping, dreaming, reading or involved in other activities demanding their complete and undivided attention. These could be individuals but also compositions of more than one person, all absorbed in the same activity. To demonstrate the effect of these (collective) states of absorption, Fried refers to descriptions like the one quoted below, describing the impact made by *Un Dessinateur d'après le Mercure de M. Pigalle* by Chardin (1753):

> How can one not be strongly moved by the truth, by the naiveté of Mr. Chardin's paintings? His figures are said not to be clever people – fine. They are not graceful – fine. But on the other hand, do they not all have their own action? Are they not completely caught up in it? Take for example the replica of his draughtsman that he has exhibited: people maintain that the heads are vague and lack precision. And yet, through this lack of precision, the attention of both figures is apparent; *one must, it seems to me, become attentive with them.* (Abbé Garrigues de Froment, quoted in Fried, 1980, p. 13, my italics)

Focalization helps me to understand the absorptive qualities of Chardin's paintings as the taking up of a position as a subject of vision. This subject of vision neither refers to an actual individual seer, nor to a subject represented in or by what is seen. Instead, it refers to a subject position implied within the visual address presented to a seer. This position mediates the relationship between an actual seer and what is seen. Focalization, then, describes the relationship between this subject position and that which is seen. Absorption describes the context in which the seer takes up the position or point of view presented to him or her, and does so without giving it a second thought. The effect achieved is in a way similar to 'taking up' the position of a character represented on stage or empathizing with a performer convincingly presenting him- or her- 'self'. The result is a sense of directness, closeness and immediacy. We are invited, momentarily, to forget the relationship between ourselves and the other we are seeing. It is as if we experience directly what the other feels, seeing the world through his or her eyes. The description of Abbé Garrigues de Froment also testifies to the relationship between such a mode of looking and notions of truth and believability. This becomes even more evident in Fried's description of a second strategy of achieving absorption, for which he cites Diderot. Diderot not only explicitly positions this truth effect in opposition to theatricality, but also presents drama as a model of an aesthetic logic (in painting and in the theatre) that guarantees the absoluteness of the picture or representation on stage relative to the seer.

Drama as a strategy of de-theatricalization

According to Diderot, nothing is more damaging to the act of persuasion than when a painter's *dramatis personae*, or *dramatis personae*, seem by virtue of the character of their actions and expressions to evince even a partial consciousness of being seen. When this happens, the figures depicted will appear mannered and false, their actions and expressions will be seen, not as natural signs of intention or emotion, but merely as grimaces – feigned posings directed towards the beholder. The work then becomes what Diderot deprecatingly calls *un théâtre*: 'an artificial construction whose too obvious designs on its audience made it repugnant to persons of taste' (Fried, 1990, p. 7).[10]

While in Diderot's writings 'le théâtral' refers to the consciousness of being beheld and is synonymous with falseness, drama, surprisingly, appears as a means to direct attention in a way that prevents theatricality. To be more precise, drama for Diderot is a means of de-theatricalizing beholding and to guarantee the absoluteness of the picture or representation on stage relative to the seer. Drama does so by constituting internal unity among the elements seen with regard to one single point of view. What Diderot calls for, as Fried points out, 'is at the one and the same time the creation of a new sort of object – the fully realized *tableau* – and the constitution of a new sort of beholder – a new 'subject' – whose innermost nature would consist precisely in the conviction of his absence from the scene of representation' (Fried, 1980, p. 104, italics in the original). Surprisingly, drama here appears as a means to do precisely what Fried earlier, in 'Art and Objecthood' dismissed as theatrical, namely the setting up of a relationship between work and viewer. But, importantly, drama appears as a means to set up this relationship in such a way that the relation itself remains obscured, as a result of which the effect is the opposite of theatricality, namely absorption.

For Diderot as for the early Fried, to convince as a true work of art is opposed to being theatrical and both associate theatricality with an address that makes the seer aware of his or her position relative to the work. But whereas Fried (in his early work) opposes this explicit relationality to the absence of such a relationship with a seer (in what he associates with 'presentness'), Diderot opposes the explicit relationality associated with theatricality with the invisibility of the relative position of the seer, that is, with the relationship going unnoticed. In Diderot it is not the absence of address and point of view that results in absorption, but that absorption results from an address that successfully obscures or directs attention away from the relationship between seer and seen, and thus provides the seer with a point of view marked by absence. In Diderot, drama thus appears as a means of *compelling conviction*: the conviction that what the seer is looking at is organized according to a logic that appears convincing as a 'law of nature,' rather than as human invention. Drama is not the only means by which to achieve this effect, absorption can be evoked in different ways using different strategies. What works and what does not work has to be understood in relation to a culturally and historically specific seer responding to the address presented.

Landscapes and points of view

The concept of point of view is central not just to Diderot's vision of painting and drama but to his epistemology. As he writes '[t]he universe, whether considered as real or as intelligible, has an infinity of points of view from which it can be represented, and the number of possible systems of human knowledge is as great as that of points of view' (*Oevres Complètes*, VIII, 211, quoted in Fried 1980, p. 216). For Diderot, the concept of intelligibility seems to entail the concept of point of view: something may be said to be intelligible only from one or another of an infinite number of points of view. This means that the claim to understand a given phenomenon, or to recognize its truth, involves accepting the responsibility, not just for the explanation itself, but also for the point of view implicit in that explanation.

Diderot makes another important observation with regard to intelligibility, vision and point of view, namely that in order to appear as truthful or convincing, the points of view implied within visions of 'how it is' must not appear too obvious. Otherwise, the effect will not be truthfulness, but artificiality, theatre. Here again the comparison with perspective is helpful. Perspectival images present a viewer with a very specific reading of what is there to be seen. Moreover, the promise given in a perspectival image is that by taking up this position, the viewer is granted an 'objective' image of how things are, independent from any specific viewing position. Moreover, although today the technique of linear perspective has lost its power to compel conviction, having become all too visible as a technique originating from a specific time and place, it appears to be much harder to acknowledge the historical character of the promise given in perspective: that of an objective world existing independently of our subjective point of view. Absorption as a strategy of persuasion plays into precisely this promise of metaphysical plenitude.

Interesting in this respect is also Diderot's account of a third strategy of absorption as used by painters of his time, one Fried terms the 'pastoral conception' of absorption. This strategy is related to landscape painting and it seems that there may well be ways of relating it to the 'textual landscapes' on the contemporary post-dramatic stage as well (I come back to this in Chapter 4). Lehmann introduces the term 'textual landscapes' to describe profound changes in aesthetic logic resulting from the transition from dramatic to post-dramatic theatre. On the post-dramatic stage, the perspective once provided by the dramatic frame is deconstructed, perverted or completely absent. As a result, the audience is granted more freedom to wander around and 'see for themselves'. Similarly, the pastoral conception of absorption results from paintings that do not position the seer by means of a clear focus. Instead, the seer is free to 'wander around'. This fiction, according to Fried, is [conspicuously at odds with the doctrine of radical exclusion of the beholder that I have argued his (Diderot's, MB) writings expound] (Fried, 1980, p. 118). Diderot describes these absorptive paintings of the third kind as if he is actually walking around in a landscape. In this description, the relationship between seer and what is seen is not denied but very much present. But who is actually present here, and in what way?

It is only after many pages of description that the reader realizes that Diderot is in fact talking about a painting. This is a critic truly absorbed in the work of art! Absorbed, yet in a different way. Diderot is not absent, as was the case with the other paintings. He seems very much present, and describes himself as looking around freely, engaging with the landscape. However, it is exactly the character of his presence and engagement that I argue is the key to the understanding of what absorption might mean here.

What is absent from Diderot's descriptions is his interaction with the painting *as painting*, as well as his physical presence as seer in relation to the painting as object seen. The 'I' in Diderot's text interacts with the landscape represented, not with the painting. This 'I' is completely absorbed into the painting, in the sense that for this 'I', the landscape does not appear as a representation but rather as a real landscape that he has entered by 'stepping inside'. This is not to say that Diderot did not know that he was looking at a painting. What I want to draw attention to is how, according to his own description, his experience of being absorbed in the painting involves the absence of awareness of the painting as representation. His being convinced by the representation of the landscape equals the seeming absence of a relationship between seer and what is seen, the absence of awareness of representation qua representation so typical of the 'subjectless' representations of the world of modern science. Also, this effect results from a very specific – yet invisible – point of view, rather than the absence of any point of view.

Here it is not the well-composed tableau that brings his desire to a halt. The effect of absorption is not achieved through fixing the seer in place as in perspectival painting or in the dramatic model of absorptive painting. On the contrary, Diderot's description testifies as to how it is in this case precisely a lack of focus that supports the illusion of unproblematic and directs access to what is there to be seen. This lack of focus seems to respond to a longing to exceed the limitations of his temporal, spatial and physical being and does so by presenting the seer with a subject-position similar to the Cartesian disembodied I/eye. The 'I' in Diderot's text seems to prove Descartes' remark that 'it is certain that I am truly distinct from my body, and I can exist without it' (Descartes, 1977, p. 235). Liberated from the bodily locus that is Diderot, he can wander around freely and experience the landscape seen in the painting as a spectacle before his own vision.

The drama of deconstruction

Diderot's observations on point of view and what convinces as truthful (re)presentation link up remarkably well with postmodern, feminist and postcolonial critique of the unified and supposedly universal point of view implied by the grand narratives. Such critiques have taught us that the deconstruction of this unitary point of view does not result in the absence of point of view or perspective per se, but rather in a multiplication of viewpoints. In this situation, focalization allows for an approach to questions of points of view as they are contained in both theatre and performance, starting from: What kind of subjective point of view is implied by what is seen? How does that which is presented address a viewer as viewer? To what kind of desires, expectations or presuppositions does this address (intentionally or unintentionally) respond? In this approach, the truth or immediateness of performance becomes a symptom rather than an ontological given; a symptom that indicates a match between the desires, expectations, presuppositions and anxieties that characterize an actual viewer and characteristics of the subject position implied by what is seen. This match, and not the absence of any received point of view, causes the jubilant experience of immediateness and directness, of being able to see it 'as it is'.

From a relational approach to visuality, the question is not whether a point of view is absent or not, but *how* focalization mediates in the relationship between the visual address presented by a painting or performance and an actual seer as subject. Seen this way, theatricality and absorption both describe the effects of such mediation, where theatricality indicates a situation in which the seer is addressed in a way that highlights (aspects of) focalization, whereas absorption indicates a situation in which such mediation goes unnoticed. From a relational point of view, neither absorption nor theatricality indicate truth in the sense of some kind of 'givenness' independent from a particular point of view. Instead, each in their own way, theatricality and absorption can alert us to elements of subjective 'investment' in what appears to us as convincing (re)presentations of 'how it is', with the world, or with someone on stage. They alert us, that is, to the relationship between our subjective desires, presuppositions, fears and anxieties and the way these are or not met by the visual address presented to us.

Looking at Forsythe's *Artifact* one more time, it seems that it might now be possible to read the 'drama' taking place between the two speaking characters as a theatrical response to Freedman's (1991) question, whether deconstruction is the enemy of theatre or its double. In *Artifact* the two speaking characters (the man and the woman) seem to be engaged in an ongoing dispute about 'how it is' and each of them presents the viewer with different positions to take up regarding what is at stake. Both the man and the woman address the audience with an invitation to take up the position of an 'I' seeing, and to see as they do. In their attempts at making sense of what happens onstage, the man and the woman represent conflicting points of view. The woman represents the Cartesian cogito with a transcendental vantage point from whence everything can be seen 'as it is'. The position of the man corresponds to a post-modern critique of these universalist pretensions. He sets out to question and deconstruct her vision. Their conflict reaches a climax in the third act when the woman starts to knock over pieces of the cardboard set showing perspectival drawings of theatrical space. In the fourth act, these pieces of cardboard are placed side by side against the back wall of the theatre, presenting a view of different perspectives on the space of the theatre, seen simultaneously and equally from varying points of view.

In this drama, language acts as a means of 'getting the picture'. The man and the woman are involved in a process of negotiating and renegotiating what is actually the case in the situation in which they find themselves. They do so similarly to how young human beings 'enter into meaning' as described by Jerome Bruner (1990). In Bruner's account, language is not just a way of naming but also a means to adjudicate different understandings of reality. Here, narrative acts as a means to negotiate between conflicting points of view. Bruner thus argues for a view of cultural meaning making as a system concerned not solely with sense and reference but with what he calls 'felicity conditions ... the conditions by which differences in meaning can be resolved by invoking mitigating circumstances that account for divergent interpretations of "reality"' (Bruner, 1990, p. 67). He demonstrates how human beings, in interacting with one another, form a sense of the canonical and ordinary as a background against which to interpret breaches in and deviations from the perceived norm. This function becomes most acute at moments of confrontation with something that does not fit in the usual and breaks with expectations. At such moments, the subject is forced to act upon what happens and actively make sense of it. At this point, the seemingly unproblematic act of 'just looking' loses its apparent naturalness and the seer has to actively produce a story or reading of what is seen, a story that will make things coherent according to his or her point of view.

In *Artifact*, the two speaking characters represent differing ways of mediating between what is happening on the stage, and the seer. Their behaviour is illustrative of the stories that make what is seen cohere, according to their respective points of view. The woman in the historical dress demonstrates the invitation to 'step inside' of the dramatic theatre. The man with his casual modern clothes and downplayed acting demonstrates a deconstruction of her highly theatrical gestures. He sets out to look for forgotten stories, and draws attention to the forgotten 'other face' or 'under side' of the theatre by opening a trapdoor in the stage floor and inspecting the wings. Both invite the audience to take up the point of view implied within the gesture of exposure demonstrated by their behaviour.[10] What counts as convincing for the man will probably not be recognized as such by the woman, and vice versa. Nevertheless, the successive positions they represent derive cohesion from the same goal that drives their quest.

The relationship proposed by *Artifact* between theatre and its deconstruction is at odds with ideas of theatrical deconstruction when deconstruction is understood to reject presenting the audience with points of view. But it is not opposed at all to Derrida's own account of deconstruction, or of how this account is inseparable from *telos* as part of how we are involved with the world that surrounds us. In *Limited Inc.* Derrida observes:

> This telos or fulfilment is constitutive of intentionality: it is part of its concept. Intentional movement tends towards this fulfilment. This is the origin of the fatality of that 'longing for metaphysical pleinitude' which, however, can also be presupposed, described, or lived without the romantic, even mystical pathos sometimes associated with these words. (Derrida, 1993, p. 121)

Longing is the movement of intentionality and is therefore a structural and ineradicable aspect of intentionality itself. It is an integral part of the way we relate to what we are confronted with. As such, it is central to the project of deconstruction. But this longing should never be confused with its cultural expressions, which are so often, as Derrida writes, tingled with romantic, even mystical pathos. How longing manifests itself in cultural articulations and what may appear as *telos*, giving direction to this longing, is culturally determined and therefore non-universal, not beyond point of view. Furthermore, what we are yearning for can never be adequately represented. Herein lies the fatality of longing as well as one of deconstruction's most important lessons.

3
Showing What Cannot Be Seen

'We can never understand a picture unless we grasp the ways in which it shows what cannot be seen' observes Mitchell (1986, p. 39). He makes this observation in the context of a discussion about the way perspective produces an image of the visible world, an image that is constructed according to a particular logic and as seen from a specific point of view. This construction is explicitly visible in perspectival drawings, in which receding lines serve to constitute unity as a result of which all elements appear as part of a meaningful totality. The point where receding lines meet (the vanishing point) mirrors the vantage point, the point from where the scene depicted is seen. The scene reaches out to the viewer, inviting him or her to occupy the vantage point. By taking up this position as implied by the construction of the image, the viewer is granted a perfect view from where everything looks the way it should. Seen from this point, the image is like a *finestra aperta*, a window opening on the world.[1]

Similarly, dramatic theatre presents its audiences with a perspective on a fictional world on stage. In his *Postdramatisches Theater*, Lehmann explains the function of this dramatic logic in terms of framing [*Rahmnung*]. This invisible logic is *teleological*. The dramatic frame provides unity and coherence in view of purpose and reason and shows the world according to invisible beliefs about world order, history and reality. These beliefs are not represented on stage in the sense that they are being made present or visible by means of theatre signs. Instead, they speak through, or are implicated within, the structure of the representation itself.

As a result of this dramatic logic, the world on stage appears as a totality, a complete world in itself that we can enter as through the window offered by perspective painting:

> The dramatic theatre, in which the scene stands for the world, can be compared to perspective: space here is both technically and mentally a window and a symbol, an analogy to the reality 'behind'. Like the *finestra aperta* presented by the Renaissance painting, it offers what might be called an equivalent to the scale of the world, a metaphorical likeness obtained though abstraction and accentuation. (Lehmann, 1999, p. 288, my translation)[2]

Drama, therefore, is not something that is framed by, for example, the proscenium arch of the box-set stage separating audience and spectacle. Rather, it is the dramatic logic that frames what is seen, inviting the spectator to see what is presented before his or her eyes as symbol of a unitary and complete world, even if what is presented is highly abstracted or consists of fragments only (Lehmann, 1999, p. 288).[3] This symbolic logic provided by drama allows the viewer to understand everything seen on stage as part of a meaningful totality existing independently from the world around it, and independent from the spectator as well.[4] The effect is what Peter Szondi (1963) has termed the 'absoluteness' of drama. Drama is 'absolute' in that it presents itself as a coherent, unitary world with an autonomous existence.[5] The audience looks into this other world of the drama from a position outside it and is invited to forget its own position as seer in relation to what is presented in order to be completely drawn into the world of drama, an effect that Szondi describes as the 'merging of subject and object'. This ambiguous relationship finds its most perfect expression in the so-called box-set stage [*Guckkastenbühne*] in which audience and performance are clearly separated. When the drama begins, the attention of the audience is directed away from its own presence in the auditorium and drawn towards or even into the world on stage. Only when the curtain closes does the audience find itself in the auditorium again (Szondi, 1963, p. 16).

Already from the end of the nineteenth century, dramatic writing and ways of staging begin to break with the rules that guarantee the absoluteness of the dramatic world and this causes radical changes in how the audience finds itself addressed by what is presented on stage. Szondi describes these transformations in terms of the emergence of an 'epic I'.[6] According to him, these transformations find their completion in the epic theatre of Brecht, which no longer aims at drawing the viewer into the world on stage but instead highlights the distance between stage and audience. In Brechtian epic theatre, the merging of subject and object typical of the drama is replaced by a subject/object opposition.

Lehmann's comparison of drama with perspective is part of a larger argument which deals with these very same transformations and with it he presents a critique of (among others) Szondi's account of them. Lehmann subscribes to Szondi's observations that, over the course of the twentieth century, the closed world of the drama becomes increasingly more open and that this causes radical changes in how the audience finds itself addressed by what is presented on stage. He agrees that in this process, epic theatre presents an important move. However, despite all the differences between epic and dramatic theatre as pointed out by Szondi, there are, according to Lehmann, also important similarities between them. Therefore, epic theatre has to be understood as still being part of the paradigm of dramatic theatre, or, even its renewal and completion or perfection (Lehmann, 1999, p. 48). Szondi fails to notice this because he fails to notice that epic theatre, like dramatic theatre, is organized as a function of telos or goal. What has changed is only the nature of the telos that holds together the vision presented on stage and gives direction to the interpretative activities of the viewer. If we take into account this telos, we see that in fact, the subject/object opposition of the epic theatre serves a purpose similar to the merging of subject and object in the dramatic theatre, namely to provide the audience with a point of view on what is presented on stage. A point of view, furthermore, that, when taken up by the spectator, conceals the subjective character of the vision presented. This is especially clear in Brechtian epic theatre, where explicitly demonstrating the difference between actor and character and explicitly showing things to be 'mere theatre' constructs an opposition of theatre and reality. This opposition helps the audience to recognize what is shown on stage as 'mere theatre', while at the same time the opposition of 'mere theatre' to something supposedly more real guides the audience in making the right interpretation.

Only later, in the post-dramatic theatre that emerges from the 1970s onward, the unifying perspective provided by dramatic and epic theatre is deconstructed and replaced by other frames, or rejected all together so that the performance appears to offer the audience more direct contact with what is present on stage. The result can be ambiguous and often confusing experiences, which Lehmann proposes to understand in terms of a political deed. Political not because of what is represented on stage, but because of the ways in which the strategies implied in the artistic logic underlying the post-dramatic theatre, draw attention to the problem of representation, of representational forms and of how these are perceived, or not. This is a micropolitics that deals with invisible patterns on the scale of bodily awareness (Lehmann, 1999, pp. 449–473).

Apprehending such post-dramatic theatre and the effects it evokes can therefore no longer be about understanding the meaning of theatrical signs as they are presented within the framework of the dramatic structure. At this point, Lehmann's account of the development from dramatic to post-dramatic theatre, and especially his comparison of drama with perspective, presents a most useful starting point for analysis. With this account, Lehmann presents a way to go beyond representational thinking in which the power of persuasion of theatre is understood to result from the way it presents an accurate representation of an absent (historical or fictional) reality. Instead, he argues, theatre and reality are better understood as parallel constructions and the success of theatre to convince is the effect of its being structured according to a logic similar to the logic at work in concepts of reality. That is, the power of persuasion of theatrical representation depends on its success in presenting us with a point of view similar to the point of view implied within what passes for reality.

However, Lehmann's account of the effect of the deconstruction of the unitary dramatic framework in terms of a multiplication of frames also raises some important questions when it comes to accounting for the subjectivity of vision as presented by the post-dramatic theatre. His description of the effect of this multiplication in terms of 'increased perceptibility' [*gesteigerte Wahrnehmbarkeit*], 'intensified presence and essence of its sensory qualities' [*gesteigertes Hier- und Sosein seiner sinnlichen Beschaffenheit*] and 'the singular lead back to itself' [*das Einzelne auf sich zurückgefürht*] seems to suggest that, no longer guided by the dramatic perspective, the spectator *is* granted more direct access to the things as they are in themselves. That, therefore, the source of the confusing experiences to which the multiplication of frames gives rise is located firmly in the thing as given 'over there' (Lehmann, 1999, p. 290). Seen in this way, the effect of the multiplication of frames would appear (at least in some respects) to equal the absence of frames. This turns Lehmann's account of perception on the post-dramatic stage into a paradox: the multiplication of frames manifests itself in the increased perceptibility of the thing in itself. This paradox becomes no less paradoxical – yet somewhat less confusing – when understood as the effect or indicator of perspective at work, rather than as the effect of deconstruction or the absence of perspective. For the paradox in Lehmann's account of post-dramatic theatre is inherent to perspective: it is the paradox that *is* perspective.

In the following, I approach this paradox starting from a closer look at the comparison of drama and perspective in Lehmann's text, and the ways in which this comparison functions as a metaphor at several levels. Through this metaphor, Lehmann embeds his account of the transformation from dramatic to post-dramatic theatre in a network of already existing meanings and invites us to 'live' the meaning of drama through these meanings.[7] Doing so, this conceptual metaphor presents us with a perspective on what is drama, what is perspective and how the two relate. My question is: whose perspective is this? Whose desires, interests and expectations are involved in an understanding of perspective as a historical and artificial construction, currently undermined and deconstructed to expose 'the singular led back to itself'? At this point, I bring in Evelyn Fox Keller's (1994) account of the paradox of scientific subjectivity for a reading of the paradox of post-dramatic subjectivity as itself a manifestation of what Damisch (discussed in Chapter 1) has termed the perspective paradigm. Acknowledging such perspective at work in the visions presented on the post-dramatic stage, I argue, is crucial to understand *how* these visions can be political, and in what way.

The paradox of post-dramatic subjectivity

In *Postdramatisches Theater*, the comparison 'drama is perspective' functions as a conceptual metaphor in at least four ways. First, the comparison draws attention to similarities between perspective painting and dramatic structure in the theatre, without saying that they are the same. As with a metaphor, the combination of similarities and differences makes the comparison productive of new insights that could not be expressed before the metaphor. The second way in which the comparison of drama and perspective functions as a metaphor is found in the way in which the comparison displaces meaning and redirects attention. As a result, formerly unseen elements of the first term come to the fore. Lehmann uses this comparison as a means of looking at dramatic theatre as if 'through different glasses', and this brings him to a new conceptualization, not only of contemporary, post-dramatic theatre but also of the relationship between theatre and drama in the past. With his comparison of drama and perspective, Lehmann ushers in a second discourse in which to place and explain the first term. This is the third way in which perspective functions as a metaphor in his text. In saying that drama is perspective, Lehmann highlights the similarities between drama and perspective in painting: both show the world according to a similar logic. Finally, there is a fourth way in which perspective functions as a metaphor in Lehmann's text and this is as a concept replacing a story. This story is a narration in the precise sense of being subjective, that is, as emanating from a particular subject.[8]

In Lehmann's argument, drama stands in the place of a historical form that is no longer convincing as a representation of the world 'as it is'. In this light, drama is revealed as artificial and historical, just as perspective in painting can no longer offer the illusion of a window opening onto the world. By making this equation, Lehmann delegates both perspective and drama to the past and opposes this past to a present in which both drama and perspective in painting are seen as being artificial. Part of Lehmann's argument, however, grounds the notion that once upon a time both dramatic structure and perspective in painting did convince as adequate representational strategies. Seen from the moment in the past that both dramatic structure and perspective were convincing as adequate representational strategies, the comparison of drama and perspective still holds, but takes on a completely different meaning. From hind sight, the metaphor 'drama is like perspective' does not equal artificiality, but the world 'as it is'. What the equation signifies, therefore, depends upon one's point of view. Furthermore, both drama and perspective can only signify the world 'as it is', as long as the artificiality of their construction remains concealed. This is what Evelyn Fox Keller has termed 'the paradox of scientific subjectivity', of which the perspectival image presents the image par excellence.

The power of a perspectival image is that it appears to be a convincing image of 'how things are' – independent of any particular observer – rather than a particular way of seeing or depicting these things. This however, is achieved through framing what is seen in a highly specific way. The practice of perspective explicitly inscribes the point of view from which an observation was made and accordingly makes evident the need to recognize the difference a change in viewpoint makes. At the same time, the image presented by the perspectival construction invites the claim that faithful obedience to specified rules will result in an image for which nature, not the individual observer, is responsible. Hence, out of its very contingency perspective extracts a new kind of veridicality: it locates in the vantage point of a particular somewhere the tacit promise of a view from nowhere.

This totalizing view is an effect of the address which is implied in the picture's composition. By taking up the position provided by the image, the viewer is granted a perfect view from where everything looks as it should. Nevertheless, it is exactly this relationship between the perspectival picture and the viewer which is obscured by the effect that is achieved, and this gives rise to a rather paradoxical situation. In Fox Keller's words:

> The picture does not offer the viewer a place within the depicted scene, but rather marks his own absence from it – an absence that is the very condition of his privilege as an observer. Perspectival realism, in other words, both depends on and guarantees the presence of an actual physical subject with a particular 'point of view' – a presence always outside the painting, yet connoted by an isomorphic absence in the painting. (Fox Keller, 1994, p. 314)

In perspective, the observer is simultaneously named by his or her location and made anonymous and disembodied by his or her adherence to specified rules. While corporeally attached to this viewpoint, he/she is released from him or herself and invested in a technique. This paradoxical relationship provides the observer of perspective with a safe position marked by absence.

This point of view as implied by the perspectival image, does not only mark a location in actual space from where the scene depicted is seen. It also marks a point of view implied in the symbolic spaces opened up by discourse.[9] The signifying systems that make up discourse pre-exist any individual. They present a 'making possible', an opening up of fields in which certain kinds of action and production are brought about. As individuals, we learn to participate in discourses and in this way we learn to make sense of the world. In participating, we learn to see the world according to the attitudes and assumptions as they are part of these discourses. These attitudes and assumptions become part of our vision of the world: a vision that we find (or do not find) reflected in the representations we are confronted with. Therefore, showing a scene from a particular point of view, an image shows more than what can actually be seen.

This intertwining of what is seen with what is not seen brings about the attractiveness and even the credibility of the perspective image. As Mitchell puts it, '[p]art of the power of perspectival illusionism was that it seemed to reveal not just the outward visible world but the very nature of the rational soul whose vision is represented' (Mitchell, 1986, p. 39). With this observation, Mitchell draws attention to perspective as a definition of the relationship between what is seen and the point of view from where it is seen as such, as well as to the complex character of this relationship as it is defined in perspectival illusionism. For, as Mitchell remarks, if there is one thing that cannot be seen in an illusionistic picture, or which at least tends to conceal itself, it is precisely its artificiality (Mitchell, 1986, p. 39). The power of an illusionistic picture is that it is convincing as an image of 'how things are' independent of any particular observer, rather than a particular way of seeing or depicting these things, while at the same time this effect depends on the viewer taking up a very specific point of view that in its turn has to remain invisible in order to produce the desired effect.

This contradiction, still clearly manifest in the composition of artificial perspective, was the problem that modern scientific subjectivity needed to solve. What solved the problem, or at least provided the illusion of having done so, is what Fox Keller has called the history of 'semiotic repositioning', in which the embodied interpreter or observer is replaced by an invisible, autonomous and virtual meta-subject. This was accomplished through a wide variety of practices including the erasure of traces of perspectival construction or framing, the development of more complicated pictorial techniques, as well as the standardization of instruments as a result of which the original observer need no longer be identified. These practices also included the replacement of the first person pronoun narrator in scientific texts with the abstract 'scientist' who could speak for every man but was 'no man'. The effect of these strategies is the disappearance of the consciousness of representation as representation, rendering perspective and point of view invisible. This invisibility is constitutive of representations of the world 'as it is' in itself.[10]

Staging vision

Fox Keller's reading of the relationship between perspective and so-called objective, seemingly perspective-less representations allows for a link to be made between deconstruction of dramatic perspective and a critique of vision as subjective. Her reading clarifies how the deconstruction of the dramatic frame can support the suggestion of a perspective-less vision of the world, of seeing things as they are 'in themselves', precisely because this deconstruction, when it involves the erasure of traces of framing, contributes to a disappearance of the consciousness of representation as representation. Fox Keller's analysis of the relationship between perspective (and its deconstruction) and modern scientific subjectivity also suggests an explanation of the political implications of the disturbing experiences that are evoked by post-dramatic theatre. Such disturbing experiences could be called political in the sense that they evoke a de-naturalization of the dominant perspective at work in conceptions of reality. In chapters to follow, I elaborate examples of such critical theatre performances that expose the perspective involved in what is 'just there to be seen' by drawing attention to the 'blind spots' in these seemingly perspective-less visions. In this chapter, I start from a comparison between two different strategies of staging plays.

In staging a play, a director inevitably relates to the perspective given within the aesthetic logic of a play. How he or she chooses to relate to this perspective will be of crucial importance for the kind of address presented to the audience. The choices made by the director will influence how the audience is invited to agree or disagree, and to sympathize with or be critical of the various characters and the course of events represented. These choices determine how the audience is invited to recognize itself in what is represented, whether the audience will be absorbed by the vision presented and convinced by it; what 'truths' it may contain in relation to the world as the audience knows it.

With his staging of *The Children of Heracles* (2002),[11] Peter Sellars explicitly claimed such relevance of this old tragedy with respect to its manner of dealing with a prominent issue in today's world. The performance was announced as 'a lucid production that turns Euripides' 2400 year old tragedy into a modern tale about asylum seekers'.[12] It attracted a lot of attention, not least because of Sellars' choice to have Heracles' unfortunate children played by real asylum seekers. Sellars himself said about this production:

> I asked the actors to be as lucid as possible. Today's world is clouded. If the situation in the world has become impossible in the political sense, if powerful answers are needed, then you have to look at the Greeks. They teach spectators to listen. In Greek tragedies, I keep seeing the same image of people in an enormous ear. Greek theatre is essentially acoustic. We end our performance with a cry for revenge. And say to the audience: it's up to you! That's what theatre can be like. Not J'accuse. But: I ask you a question. I demand space for reflection. (Sellars in an interview in the French paper *Libération*, 20 November, 2002, quoted in the Holland Festival Program, see Zonneveld, 2004, p. 168)

Concerned with the problems of asylum seekers, Peter Sellars looks at the Greeks and finds his answer in Euripides' *The Children of Heracles*. With his choice to cast real asylum seekers as the children of Heracles, he stresses the connection between this old story and what happens to asylum seekers today. Sellars' commitment to the situation in the world today not only motivates his choices for particular topics or plays, but also inspires ways of working that demand extensive personal involvement. During the preparations for *The Children of Heracles,* Sellars personally visited centres for refugees in the different countries where the performance was to be shown. During these visits, he invited some of the children to participate in his show. He also invited adult asylum seekers, former asylum seekers, local politicians, opinion makers, writers etc. Prior to every night's performance, a debate was held between several of these guests, followed by an introduction of the children about to perform in *The Children of Heracles*.

In this pre-programme, the presence of real asylum seekers confronted the audience with their reality in a way that numbers or reports in newspapers and on television could never do. Their presence and their stories helped to complicate the audience's outlook on reality by confronting us, the others to whom this performance was addressed, with the limitations to our view of the world. In the performance, however, the being real of the asylum seekers worked in a very different way and to a very different end. Casting real asylum seekers as the fictional children of Heracles (and stressing their realness throughout the publicity and the pre-programme) rather than confirming the reality of asylum seekers in our society, instead confirmed the reality and actuality of the view presented by the play. 'They are real, they are refugees too, so this must be their story, this must be how it is', is what the gesture of exposure presented by the play seemed to say. This realness of the asylum seekers worked to obscure the fact that what was shown on stage is a drama told from a specific point of view, and this point of view is not that of the refugees.

For, unlike what the title *The Children of Heracles* might suggest, this play is not so much about asylum seekers as it is about those who are landed with them. This tragedy tells the story of old Iolaos who, together with the children of his former brother in arms Heracles, is on the run for King Euristheus. Euristheus wants to kill the children of Heracles because he fears they are a threat to his throne. Iolaos and the children have asked for asylum in several cities already, but time and again these cities refused their protection when put under pressure by the powerful Euristheus. When the play begins, they are in Marathon, in the kingdom of Attika, where they receive a warm welcome until, again, Euristheus' messenger Kopreus appears and demands their extradition. First, the rulers of Attika resist. It is their moral obligation to protect the refugees, they argue. However, their resistance begins to crumble when Euristheus' army advances and war and destruction are at hand. At that moment, an oracle tells them that the gods demand the sacrifice of a virgin child in exchange for their support in the war. What to do in this dilemma? How to choose between physical safety and upholding moral principles at the cost of literal sacrifice of one's own flesh and blood? Luckily, one of Heracles' daughters decides to sacrifice herself. With all possible honours she is killed, after which the old Iolaos transforms into a youthful war hero who leads the army to victory. In the end all is good.

Casting real asylum seekers as Heracles' poor children does not change the fact that Euripides' tragedy is a story in which asylum seekers are represented from the point of view of those being confronted with them. Casting real asylum seekers works as a strategy of de-theatricalization that obscures how the narrative logic of this story is constructed to respond to the anxieties of a viewer who is struggling in the face of this confrontation; one who feels morally obliged to help while simultaneously experiencing this obligation as a burden, even as unfair, especially when helping involves taking a risk. More than that, the play confirms this viewer in the comfortable conviction that he or she is to be pitied, that confrontation with such a situation is indeed a burden, even a tragic fate. Sellar's staging does nothing to question or criticize this perspective; it does the contrary. One salient detail is Sellars' choice to have Iolaos played not by an asylum seeker but by a professional actor with Western looks. Within the story, as represented in Euripides' play, the position of Iolaos is ambiguous. Strictly speaking, he belongs to the group of refugees. Although not himself one of the children of Heracles, he is on the run with them. Nevertheless, the way he speaks about the children suggests that he takes a certain distance from them. He acts as a kind of guardian who feels responsible for them but does not understand himself as being one of them. Casting a professional actor with Western looks as Iolaos increases the distance between Iolaos and the group of refugees. Doing so, Sellars visually shifts the border between those who are asylum seekers and those who are not, and supports the suggestion that Iolaos belongs to the others. Moreover, since Iolaos is the spokesman of the refugees and the only one of them who speaks, this 'repositioning' of Iolaos also turns the play from a dialogue with asylum seekers (Iolaos being one of them) into a dialogue about asylum seekers. In this dialogue, the asylum seekers do not partake.

This could have resulted in an interesting reading of the play, provided that Sellars' staging invited a critical attitude towards the way in which the asylum seekers are silenced and excluded from the debates about them, and an equally critical attitude towards the verbal haggling with which these well-meaning characters try to make their beautiful moral principles meet their practical interests. In this way, the performance could have held up a mirror to its audience, confronting it with a historical example, the argument with which we all too often try to salve our consciousness as we walk away from responsibilities. But this is not what happens in Sellars' performance. Whereas Euripides' text might have helped to raise questions about contemporary attitudes towards asylum seekers, Sellars instead reduced the asylum seekers to silent extras in a performance that first and foremost invited spectators to wallow in the 'tragic fate' of those to whom the asylum seekers direct their plea for help. Time and again, the performance stresses the helplessness of Iolaos (in a wheelchair) and the rulers of Attika, confronted as they are with the 'inevitability' of fate. Instrumental here is the way Kopreus (modelled after Condoleeza Rice) is presented as the representative of absolute, uncompromizing, totally unreasonable Evil, while in fact Kopreus' text, however much we may disagree with what it argues for, consists of arguments that are not that unreasonable at all, and are at points remarkably similar to the arguments with which the other characters try to talk themselves out of their moral dilemmas. Equally persuasive is the way Iolaos is staged as the emotional centre of the performance. Much is made of his lamentations while the emotions of the children remain silent. Visually, the children are in the margin of the stage and dramatically they do not play a part. This becomes especially clear at the only moment in the play one of the children does speak, the crucial offering of self-sacrifice by one of the daughters of Heracles.

In the play, this offer is a true *deus ex machina*. The situation is hopeless, and suddenly this daughter appears, gives a short speech in which she argues that offering herself for her brothers and Iolaos is her moral duty, and so it happens. Until that moment, she had been offstage, hiding, as we have been told, in the temple. Apart from the short monologue, the play gives no attention whatsoever to her motivations, or her emotions, to her struggle with the situation she is confronted with, or her process of decision making. She appears out of nowhere, offers herself and dies. Everybody accepts this offer without hesitation or question, and Sellars' staging confirms this narrative logic. The daughter of Heracles (also performed by a professional actress) accepts her fate without much emotional comment from anyone involved (including herself), while at the same time much is being made of the emotions of the non-refugees, all of whom are grieving that things have to go this way. Attention is focused on the respect with which she gets sacrificed, and away from anything that could tell us something about her perspective on the events. Everybody grants her all possible honours, while at the same time nobody seems to be really moved by what is happening to her. The only emotional reactions are from Iolaos, but again, these emotions concern first and foremost himself. She is going to die to save him and the others, but when she asks him to stand beside her when she faces the knife, he refuses this last wish, saying: 'I couldn't bear it, standing beside you, watching you die' (Euripides, 1981, p. 49). Poor Iolaos, fate is so hard on him.

Seeing Sellars' *The Children of Heracles* was an uncomfortable experience, and not because this performance opened a 'space for reflection', as Sellars himself claims. On the contrary, it seemed that Sellars' focus on the Greek theatre as being 'essentially acoustic' has closed his eyes for the kind of vision his performance presents to its audience, and to how this vision prevented the audience from listening to what is actually being said. We are all victims at the mercy of a cruel fate, is what Sellars' performance suggested, thus directing attention away from the questions of responsibility that Euripides' text could have raised, as well as from the radically different position the two parties (refugees and others) occupy with respect to this 'tragic fate'. The presence of real asylum seekers obscures the complexities of the problems addressed, as well as the perspective implied in the way in which this problem is represented in the play. Why, for example, does Euripides use so many words to explain the motivations, thoughts and feelings of Iolaos and the rulers of Attika, while Heracles' daughter's choice to sacrifice herself apparently needs hardly any explanation at all? Furthermore, the emphasis put on the realness of the asylum seekers representing the children of Heracles and the personal details with which they are introduced, only obscures the fact that the position of the children of Heracles within this story cannot be understood from who these asylum seekers representing them really are; nor can the reality of the asylum seekers be understood from the way in which the children of Heracles are represented in Euripides' play. Their position in the play has nothing to do with who the asylum seekers really are, but is the product of ancient Western imagination as expressed by Euripides. The conflation this performance invites – that of the position of the real asylum seekers with that of the children of Heracles – is also an invitation to take claims this representation makes about asylum seekers as reality. The presence of real asylum seekers and the rhetoric surrounding the performance suggests that the spectator can see them as they are in themselves. In the performance however, their presence only confirms the vision of those who silence and marginalize them. A powerful answer indeed.

A few inquiries

A radically different approach was taken by Gerardjan Rijnders in his staging of another story from ancient Greece, which is the tragedy of Antigone in the opera version of Tommasio Traetta (*Antigona*, 1772).[13] Whereas Sellars invites identification with the characters of *The Children of Heracles* as the powerless victims of the situation they find themselves in, Rijnders's staging illuminates how what convinces as good and evil, right and wrong, is bound up with position, power and desire. He shows, not only how this affects the choices made by the characters within the play, but also how this intertwining of morals preached by the play and the narrative logic of the play (including the conclusion it works towards) can be seen to respond to a particular (historical) point of view with respect to how things are and should be; how they correspond to a vision of reality and how this vision of reality is at work in how we see what we see. Telling in this respect is a remark by Rijnders (quoted in the programme brochure) concerning the rather surprising happy end of Traetta's *Antigona*:

> Traetta composed Antigona for Russian Empress Catharina the Great. Needless to say that confronting her with the fact that kings can be mistaken was out of the question. In the eighteenth century, the authorities were always right and therefore, in this opera, Creon had to repent. In our staging, we show that a few queries are in order here. It is a happy ending because everybody stays alive, but how happy is this ending really? (Rijnders, quoted in the programme brochure, my translation)

Like the sacrifice of the daughter of Heracles, Creon's repentance is a *deus ex machina*. In his staging, Rijnders does not take this ending for granted, but stresses its artificiality in a way that highlights the relationship between this ending and the demands of the kind of seer that this ending addresses. Totally unexpected, Creon, who had been the bad guy from the beginning, sees the light. The other characters accept this change as a matter of course, without asking questions or showing any emotion. The mise-en-scene and movements that had from the beginning been highly stylized now become mechanical. The members of the choir move like mechanical puppets while Antigona and Ermone are put into eighteenth-century wigs, as if to stress the fact that this ending was constructed to function like a mirror in which the eighteenth-century audience would want to recognize itself. The result is a most depressing happy ending that gives food for thought.

Instead of sweeping the audience along with the tragic fate of characters represented on stage, Rijnders's staging of *Antigona* presents an invitation to question what is presented by the story as inevitable and necessary. For example, what should we make of the curious fact that in this tragedy about a sister who cannot accept that one of her brothers is getting a decent burial and the other is not, there is not a moment Creon's decision that it should be Eteocles who is buried and not Polineikes is contested? Antigone argues that as a sister she has the obligation to bury Polineikes as well, whatever he has done. But no one, not even Antigone, questions Creon's decision to appoint Eteocles as the good guy and Polineikes as the bad guy. This is quite striking if one takes into account the history that preceded the death of the brothers. After the death of their father Oidipous, they agreed to rule on an alternating basis. Eteocles is the first one, but he refuses to hand over the throne to Polineikes when it is his turn. Polineikes then assembles an army and marches against Thebe, his own city, and his brother Eteocles. In the fight that follows, both die.

'When hegemony is under siege, when the imagined political landscape is under attack, there is little tolerance for complexity of meaning. But complexity is just what the diverse multitude in a global public sphere demands' observes Susan Buck-Morss (Buck-Morss, 2003, p. 27). *Antigona* represents such a moment of hegemony under siege. According to their former agreement, Polineikes rightfully claims the throne. Yet, Eteocles, once in power, refuses to acknowledge what they agreed upon. So, actually, Eteocles is the bad guy, the one that causes the problem that turns Polineikes into the enemy bringing war to Thebe, his own city. But Eteocles is also the ruler of Thebe (how can he be wrong) and Polineikes the captain of the hostile army. Once both brothers are dead, Creon decides to ignore the history that would complicate the question of right and wrong, as well as his own position in relation to this history (a strategy not unlike that of many political leaders in our times as well). His decision reinstalls a clear cut distinction between right and wrong by means of a brutal simplification that denies history. You are either with us or against us.

Rijnders's staging visually comments on this simplification as it is part of the vision presented by the opera. At the beginning of the performance, both brothers are represented by their clothes: two empty suits laying side by side on a platform. The only difference between the two is the crown placed on one of them. In the first scene, Creon puts the crown on his own head and then declares that it will be Eteocles who will be buried and Polineikes will not. Without the crown, the two suits representing the brothers are exactly the same, stressing the arbitrariness of Creon's decision. At the same time, the direct relationship between Creon's appropriation of the crown and his choice for Eteocles and against Polineikes suggests that his choice of Eteocles is (at least partly) motivated by his own identification with the position of power.

Although Rijnders avoided direct reference to actuality, his way of staging nevertheless invited reflections that are most relevant with respect to the questions and complications we find ourselves confronted with today, albeit in a totally different way than Sellars' *The Children of Heracles*. Instead of presenting *Antigona* as the answer to today's questions, Rijnders highly stylized and explicitly theatrical performance illuminated how Traetta's representation of the story of *Antigona* is constructed to present an answer, and how this answer implies a perspective that responds to a particular point of view. Rijnders demonstrates how this old story of *Antigona* gains power of expression when this point of view is not taken for granted but put in perspective, thus allowing for a step towards an ethics of vision that illuminates how norms and values are the product of culture and dependant upon point of view.

While Sellars brings in real asylum seekers to stress the reality of the story told on stage, Rijnders's staging only refers to other representations. The scenery 'opens up' to a world 'outside' that consists solely of historical paintings of cityscapes and landscapes (Figures 3 and 4). The performance, however, does not aim at a reconstruction of this historical reality. The overall design is sober, stylized and clearly early twenty-first century. Within this setting, the meaning of the paintings is symbolic. By showing the world 'outside' as one consisting of representations, Rijnders points attention to the fact that our only way of knowing reality is through the cultural and historical perspective that is part of our way of looking at it, and that this looking is always mediated by representations. The real world 'outside' is perspectival as well and this perspective inevitably structures our way of representing and knowing the world. Furthermore, the eighteenth-century paintings remind the audience of the historical character of what they are watching. That is, these paintings serve as a reminder of how reality as represented within *Antigona* implies a historically and culturally specific point of view.

Rijnders used a similar strategy in many of his stagings of historical plays. In *Penthesilea* (1991),[14] for example, the lethal battle for love between the Greeks and Amazons was set in a library. The set, as well as the elegant early-nineteenth-century clothing, was reminiscent of von Kleist's own historical period, highlighting von Kleist's frame of reference from which the mythological story in the play is presented. The design of the library was remarkably similar to that of the Teatro

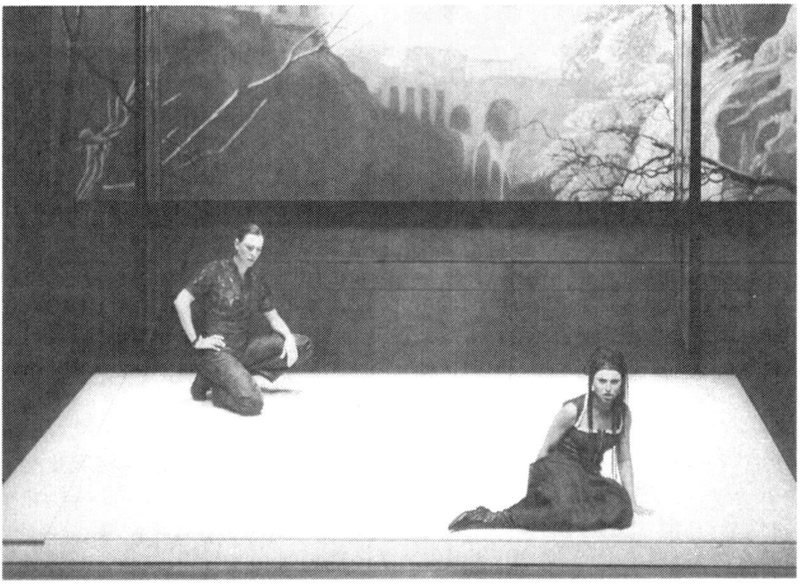

Figure 3 Photograph by Herman Sorgeloos of Rafaella Milanesi and Maartje de Lint in *Antigona* by Gerardjan Rijnders. Reproduced with permission of the photographer.

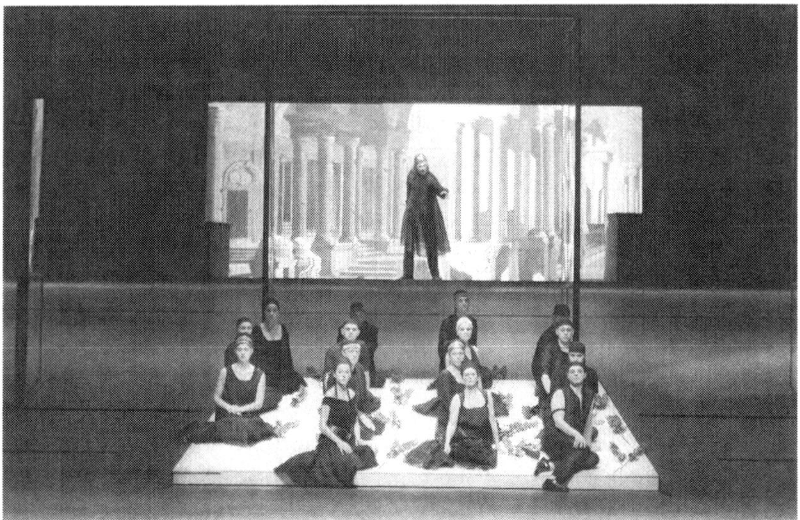

Figure 4 Photograph by Herman Sorgeloos of the chorus in *Antigona* by Gerardjan Rijnders. Reproduced with permission of the photographer.

Olimpico, the famous Renaissance attempt at reconstructing the classical theatre. Hence, instead of suggesting a transparent window to the classical world of Achilles and Penthesilea, Rijnders's staging alerted the audience to the ways in which its relation to this history is mediated through a complex web of historical references and a superposition of visions from various historical periods. A similar strategy was used in *Andromache* (1990),[15] in which décor and costumes referred both to antiquity and to Racine's day, where Louis XIV peruques were worn with classicist versions of antique costumes.

Rijnders's stagings of historical drama draw attention to the relationship between visions of historical periods as presented in plays, as well as historic and culturally specific points of view. They make audiences aware of how visions of historical times and events always imply such points of view and how these historical worlds are made present in relation to these points of view. Rijnders's stagings also draw attention to the ways in which these points of view mediate the relationship between a contemporary viewer and visions of historical worlds represented on stage. This relationship between historical visions and contemporary viewers was represented on stage in Rijnders's *Klaagliederen* (1994)[16] based on Jeremiah's *Lamentations* in the Bible (which performance is also mentioned by Lehmann as an example of post-dramatic theatre; Lehmann, 1999, pp. 233, 400).

Klaagliederen evoked memories of Rembrandt's paintings of biblical subjects. In heavy baroque costumes and with slow, stylized movements the actors moved between the pillars of a 'cathedral'. By using a seventeenth-century translation of the Bible, the performance explicitly generated a sense of historical distance. In the midst of all of this, a man, dressed in a late-twentieth-century suit, could be seen wandering about. Although he tried to connect with the other actors and to partake in the ritual they seemed to be performing, his efforts were all in vain and he remained an outsider. At one particular moment, he looks at an angel, sitting motionlessly in a corner of the 'cathedral' and reaches out to touch its wing. The movement stands out because of its resemblance to the famous image in Michelangelo's painting in the Sistine Chapel: the *Creation of Adam* (1508–1512), which portrays God reaching out his hand to touch Adam's fingertips and give him life. In *Klaagliederen* it is the human being that gives life to the historical supernatural creature, bringing it to life with his look.

The image presented at this moment echoes Michelangelo's famous painting of this crucial moment in the Judeo-Christian story of Creation, a moment of origin par excellence, painted on the ceiling of a building at the centre of the Catholic universe. Yet it is neither its status as origin, nor its central position, that makes it such a powerful image today. People who 'recognise' the image – that is, people for whom it stands out as meaningful or recognizable – may not even be aware of its origin. The image has a history of its own, endlessly reproduced as it is on postcards and calendars, to shower curtains and cookie tins. But its history also consists of the ways in which this image has been used and reused in other images, from cartoons to masterpieces, including *Klaagliederen*. The image has become part of our visual unconscious, the repertoire of images through which we 'see' the world, including the ways we 'see' historical times.

The various elements that together made up Rijnders's version of *Klaagliederen* were kept together by a kind of logic that did not result from an origin shared by them. Rather, what made this look like a convincing representation of biblical history, was the ways in which the language spoken, the story told, the appearance of the characters and the set design all referred to the same history of imagining the biblical past. Just as important to an understanding of how vision 'works' here as the intertextual relationships between the biblical stories and Rembrandt's paintings is *how* these are related to a subjective point of view that is part of this history as well.[17]

This subjective point of view mediates the relationship between us, the audience, and the vision presented on stage. In *Klaagliederen* this happens quite literally. The man in the contemporary suit represents the attempts of late-twentieth-century viewers to connect with what is shown on stage. The moment at which the image recalling Michelangelo's *Creation of Adam* is crystallized, the distance between history and presence is suddenly somehow bridged. This is not because there is no historical perspective involved here, but rather because the familiarity of the image in today's culture tends to naturalize the perspective involved. This naturalization produces what Norman Bryson in his analysis of perspective calls a 'presence effect'. This presence-effect – and not the correspondences between image and the historical reality depicted – turns the image into a convincing representation of 'how it is' (Bryson, 1983, p. 94).

In his analysis, Bryson argues against the persistent notion of the perspectival painting as a copy of reality and linear perspective as a technique developed to capture a natural perspective, perceived in the world seen around us. What is suppressed by such an account of painting as the record of a perception is the social character of the image, and its reality as a sign. What is overlooked, as Bryson points out, is that the world 'as it looks' is itself subject to convention (Bryson, 1983, p. xii). This perspective cannot be taken away in order to expose the world 'in itself' behind it, but this perspective can be denaturalized in order to expose its workings. And this indeed produces ambiguous and confusing effects since this exposure and denaturalization undermines the safe position of the spectator marked by absence.

In *Klaagliederen* this happens, for example, when God Himself makes his appearance. He is accompanied by the angel and by a young woman. All three are dressed in Baroque costumes that recall images of God and other heavenly creatures in religious paintings and sculptures. God takes a few steps in the direction of the audience and then begins to speak: he condemns the people of Jerusalem for their sins and announces that he will punish them. So far, this staging of God's apparition confirms traditional representations of the scene, just as the words spoken confirm the destruction of Jerusalem as a punishment from God. Yet there are some remarkable differences as well.

In *Klaagliederen,* God and his companions do not materialize in the usual guise as Whites. Instead, these three heavenly creatures are people of colour. Furthermore, the actor playing God was familiar to (at least part of) the audience. The actor, Michael Matthews, was also well known as a theatre maker. In his own performances, he had made no secret of the fact that he was gay and suffering from AIDS. This materialization of God in the guise of Matthews, or vice versa, is not only fundamentally different from conventional representations of God and draws attention to the presuppositions at work in conventional ways of envisaging Him; of the ways that conventional visualizations play to audiences' desires, expectations and presuppositions, but also how the theatre as a 'vision machine' can be used in a critical means of playing these presuppositions, exposing them and subverting them.

Hallucinations

With their elaborate sets and costumes, Rijnders's stagings of Corneille, Racine, von Kleist and the *Lamentations,* bring to mind historical forms of illusionistic theatre, the kind of theatre that presented the audience a vision into another world behind the proscenium arch. This is another world that we, the audience, can look into or even lose 'ourselves' in while our actual presence remains literally in the dark. In Rijnders's stagings of these historical plays however, dramatic perspective is not deconstructed or broken but rather highlighted. The theatrical frame is not absent or undermined by strategies of deconstruction and de-theatricalization, but exaggerated. In Rijnders's staging of Corneille's *L'Illusion Comique* (1997),[18] the frame is literally multiplied. This performance featured not one proscenium but three. One of the three arches could be opened and closed like a diaphragm, emphasizing the play-within-a-play scenes. This is theatre, there can be no doubt about that, yet the effect was all but reassuring.

The multiplication of frames, as Lehmann observes, cancels out the operation of one single frame and draws attention to the problem of representation, of representational forms and how they are perceived. Rijnders's stagings of historical plays show this problem of representation to be a problem of the relationship between inside and outside: the relation between inside and outside of the frame, as well as of the relation between subjective interior and cultural historical outside. His performances demonstrate how frames are constitutive of the very opposition of inside and outside that grants the spectator of the dramatic theatre the illusion of a safe position as outsider who can see everything without getting really involved. They also show that the multiplication of the frame does not so much increase perceptibility of things as they are 'in themselves', as point to the inevitable intertwining of what is seen with the expectations, desires and presuppositions that are part of the subjective point of view from where they are seen.

In several of his works, Rijnders explicitly engaged with perspectival vision and its relation to the theatre as an apparatus of vision. His *Dark Lady* (1999)[19] for example used the full, nearly fifty meters of depth of the stage of the Amsterdam Muziektheater. This depth was further exaggerated by means of techniques that brought to mind perspectival stage design. Vision as presented by such perspectival design is based on what Bryson terms the differentiation of the corporeal glance from the idealized, disembodied and monocular glance. Perspectivalism turns the one seeing into a detached spectator rather than an actor in the visible world and thus supports the illusion of the detached disembodied eye/I 'just looking'. This reduction involves an invisible logic according to which the perceptual field appears as a homogeneous, regularly ordered space, there to be duplicated by the extension of a grid-like network of co-ordinates. This grid that is space is opposed to an extension-less point of view from which it seems to emanate (Bryson, 1983, pp. 94–96). In *Dark Lady*, a floor pattern of broad bands of alternating colour parallel the border of the stage, combined with a seemingly endless series of wings receding into the depth of the stage to create the impression of a grid. However, instead of firmly positioning the spectator, the highly exaggerated perspective created a dazzling effect of undecidable depth.

In this depth characters in spectacular costumes could be seen wandering around, appearing and disappearing again in the wings, with no clear purpose or reason. So rather than increasing perceptibility, the exaggerated perspective turned seeing into an experience akin to hallucination.

A similar strategy was used by Rijnders in his staging of Heiner Müller's *Kwartett* (1999).[20] Here again, he literally doubled the proscenium arch. Covered with black velvet, this double diaphragm imitated the double proscenium of Wagner's *Festspielhaus* in Bayreuth. In *Kwartett*, not only the proscenium, but the entire stage was hung with black velvet. The performance took place in near darkness. Initially the audience could barely see the whitish shapes of the two actors who were naked during most of the performance. As Crary (1999) points out, in Wagner's *Festspielhaus*, the multiplication of the proscenium arches, combined with the extreme darkness of the theatre, was intended to detach the illuminated stage from any legible relation to the rest of the opera house. The aim was to increase visibility and to support the illusion of detached spectatorship for which perspective is a conceptual metaphor (Crary, 1999, pp. 247–257). Rijnders exaggerated Wagner's strategy to the point where it began to undermine rather than support the intended illusion. Instead of suggesting a glimpse into a magical world behind the frame, framing here undermined the apparent certainty about the border between what is seen and the projections of those who are seeing it, leaving it up to them to decide what was actually there and what was the product of their own fantasies, desires, fears and imagination. Rijnders's staging thus not only presents an excellent visual translation of a central theme of Müller's play, but also one that turns this play into a reflection on theatre itself as a vision machine. His staging points to the ways in which the spectator is always involved in what is seen, even in those instances where what is seen is stripped bare to its essential nakedness, like the bodies of the actors in *Kwartett*. This world is therefore multiple 'as it is' and does not exist 'in itself', independently from points of view. And herein lie the political stakes at issue, which require urgent attention perhaps now more than ever.

4
Walking the Landscape Stage

> Look, I never liked that story either, about man set in a cave somewhere, contemplating the shadow of reality. If you ask me, the story stinks! When I think about it, I only see long-haired cavemen with filthy beards drawing animals on walls, I don't want to have anything to do with that garbage. Or with Diogenes who is supposed to have spent the whole day sitting in a barrel discussing the meaning of life with all those wise men. God, that must have been hell! I reckon that it all still makes sense, even without the cave and Diogenes. Truth is like a flash of light that is always behind you. And if this is the case, then what do you see in front of you? Nothing but your own shadow. You see, you just can't avoid getting in the way of the image of reality! (Jeroen van den Berg: *Sailors on a Bus*)[1]

The people in Plato's parable of the cave are tied to their places, watching shadows on a wall. They do not see that that which they perceive to be in front of them is actually located above and behind their heads. They mistake the shadowy projections for the real thing. Freddie Rokem calls Plato's story 'the first full fledged literary representation of spectators in the history of Western culture' (Rokem, 2003, p. 14). These are spectators who cannot choose for themselves where to look. Fixed in a given position, in the dark, isolated from what he or she sees, the spectator can only gape passively at the spectacle taking place.

Plato's parable of the cave is one of the founding myths of Western metaphysics and one of the seminal sources in the Western canon for the underlying conception of sight which is constitutive of the subject/object dualism typical of Greek and later Western metaphysics. Plato's cave, if understood as a 'vision machine', allows the observer to avoid direct engagement with the object of his gaze. The gain, Hans Jonas writes, 'is the concept of objectivity, of the thing as it is in itself as distinct from the thing as it affects me, and from this distinction arises the whole idea of theoria and theoretical truth'.[2] What gets lost here, as Martin Jay observes in his reading of Jonas, is a clear sense of causality, as the constitutive link between subject and object is suppressed or forgotten (Jay, 1993, p. 25).

Throughout the twentieth century such a spectator position has been subject to much critique in avant-garde theatre. In her article 'Discovering the Spectator' Erika Fischer-Lichte speaks of a fundamental change in the structure of theatrical communication. From the end of the eighteenth century, theatrical focus was centred on the characters on stage and the internal communication between them. In the avant-garde theatre of the early twentieth century, however, the focus of interest shifted to the external system of theatrical communication: the relationship between stage and auditorium (Fischer-Lichte, 1997, p. 41). Lehmann makes a similar observation with respect to post-dramatic theatre when he remarks that:

> The new textuality of the theatre (or the textuality of the new theatre) produces a peculiar shift of axis: it does not probe the traditional centre of theatre-discourse, the dialogue with its implications of dialectic order and intersubjectivity. However, the dialogue does not simply vanish. ... while the dialogue on stage is fading, dialogue returns with a new emphasis between stage and audience. (Lehmann, 1997, p. 58)

Fischer-Lichte explains this 'shift in axis' in terms of opening up new spaces, aimed at a new kind of spectator involvement in what is seen on stage. The spaces opened up by avant-garde experiments in the theatre no longer present the spectator with the organized frame of the box-set stage. Instead, they aim at creating a new kind of 'unity' of actors and spectators and even at transforming the spectators into 'new' beings:

> [t]o transport them into a state of 'intoxication' or 'trance' to liberate their creativity and develop it or simply to shock them. In either case, the powers of vision and hearing as practised in the bourgeois theatre, and, moreover, in the whole of Western culture for more than three hundred years, were destroyed. (Fischer-Lichte, 1997, p. 51)

The spectators lose their safe places in the dark auditorium. They lose those places – marked by absence – from whence they might merely observe a dramatic world unfolding 'over there'. Instead, spectators are addressed more directly and made aware of the fact that the theatrical event is taking place here and now. This 'here and now-ness' is further stressed by strategies of deconstruction that aim at breaking open the coherent world represented on stage in order to show what is *really* there: actors, objects and a theatrical machinery.

These stratagems, developed by early-avant-garde theatre makers, were developed and greatly radicalized on the postmodern stage. Fischer-Lichte argues, however, that postmodern theatre differs in important respects from its avant-garde ancestors. Stratagems carried out by avant-gardists were intended to reach out and transform the audience, and functioned solely as instruments bringing art closer to life: to shock the spectators, or to liberate spectators' creative potential and transform them into 'new' beings. In the post-modern theatre, on the other hand, spectators are not turned into 'material' for specific purposes from which, with the means available to theatre, 'actants' might be produced. Instead, spectators are 'given back their right to spectate' (Fischer-Lichte, 1997, p. 57). Fisher-Lichte understands this right in terms of increased freedom on the part of the spectator to attribute meaning at will:

> Post-modern theatre elevates the spectators to absolute masters of possible semiosis without, at the same time, pursuing any other ultimate goal. The spectators are free to associate everything with anything and to extract their own semiosis without restriction and at will, or even to refuse to attribute any meaning at all and simply experience the objects presented to them in their concrete being. (Fischer-Lichte, 1997, pp. 57–58)

Postmodern theatre turns the spectator into a 'master of possible semiosis'. Freed from his or her fixed position and no longer forced to see in one way rather than the other, the spectator is granted the freedom to see and to give meaning at will – or not to attribute any meaning at all – to the experience there to be apprehended.

Have we finally been set free to see the light?

Fischer-Lichte's account of spectatorship in postmodern theatre testifies to an awareness of how new strategies of staging leave space for subjective aspects of vision. Instead of directing attention in such a way as to support the suggestion that we are looking at a world existing independent of us, seers, strategies of staging now stress the relationship between performance and audience and invite more active involvement. At the same time, however, Fischer-Lichte's account of the spectator as a 'master of possible semiosis' assigning meaning at will, or not at all, suggests a denial of precisely such an understanding of vision as necessarily always subjective. The notions of 'subject' and 'subjectivity' imply a critique of the idea of the individual as an agent free to assign meaning. They also imply a critique of the assumption that it would be possible to choose to experience things 'as they are' in themselves, in full measure of their ontology. The notions of 'subject' and 'subjectivity' point to the inevitable relationships with cultural surroundings and help us to conceive of human reality, the world 'as it is', as a particular construction, the product of signifying activities which are both culturally specific and generally unconscious. Set free to see as we want to see, we are less free than we would like to think we are. This state of being bound by and to a subjective position is what gets obscured in Fischer-Lichte's definition of the spectator as a 'master of possible semiosis'.

What is needed to account for the subjectivity of vision is to expose the relationship between seer and seen and how the 'landscapes' on the contemporary stage mediate in this relationship? How does the address presented by these 'landscapes' relate to the desires, presuppositions and expectations of a culturally and historically specific seer? Who is this seer? What can the landscapes on stage tell us about this seer and about how vision takes place as a result of the interaction between seer and the onstage aesthetic logic? In what follows, I explore this relationship, starting from what might be called a postmodern variation on Diderot's description of a picture of a landscape (discussed in Chapter 2), and Heiner Müller's *Bildbeschreibung* (1984; translated in English as *Explosion of a Memory/Description of a Picture*, 1989). In 'Discovering the Spectator', Fischer-Lichte refers to *Bildbeschreibung* as 'the most appropriate expression of the post-modern awareness that looking is a creative act': in *Bildbeschreibung* 'the reception of the image is executed as the production of a text – reception *is* production, looking on *is* acting' (1997, p. 58, italics in the original). In Müller's text, looking is indeed presented as a creative act, as a 'reading', in the sense of producing meaning. However, to what extent the one seeing can be understood as the master of possible semiosis seems to be precisely the question at hand.

Getting lost

As the title indicates, *Bildbeschreibung* consists of the description of a picture. From this description the reader is led to understand that the picture presents a landscape. *Bildbeschreibung* presents it reader with precisely that: it consists of a description [*Beschreibung*] of a picture [*Bild*] of a landscape. This description offers an account of how someone seeing actually makes sense of what is seen, or at least, tries to do so by isolating and contemplating individual visual elements and combines them to make a story. *Bildbeschreibung* reads as another example of language used to actively 'get the picture' by means of a story that will make what is seen cohere according to a point of view, in a way similar to the two speaking characters in *Artifact* (Chapter 2). As much as the reading provided by *Bildbeschreibung* of what is there to be seen testifies to the freedom to connect and give meaning, it also expresses how cultural habits and conventional patterns mediate in practices of reading what (we think) we see.

The description starts with a landscape and clouds, and then it begins to zoom into the details: a bird, a house, a woman, a man. The fate of the woman in this narrative seems to confirm the gendered-ness of such a reading. The woman is an object to be looked at and does not look back. She rises up from the earth and seems to be part of the land that is violently traversed by the man. She is part of the space that is turned into narrative. At the same time, this reading is presented as a reading that does not hold together. The woman will not be destroyed. She reappears time and again, turning violation into a dazzling repetition. The thread of narrative becomes entangled, spread out like a spider's web, stretched and shaken by a battle of all against all. It appears to be impossible to fit the woman into a coherent story with a beginning, a middle and – most importantly – an end. *Getting the picture* also appears to be impossible, since Müller's text presents its readers with an account of interaction between the one seeing and what is seen in which the one seeing gets lost in the landscape of the picture.

Just as the reading of the picture does not express freedom and mastery over possible semiosis, getting lost does not appear to result in freedom and mastery. Rather, the text testifies to obsessiveness, to getting stuck and to the desire to escape. The verbalization does not and will not completely match the picture seen. The seer gets caught in obsessive loops and is lost in partial descriptions that shift while he or she speaks on. Finally, the speaking 'I' seems to lose all sense of distance between itself and what is seen and collapses into the picture, saying:

> who OR WHAT inquires about the picture, TO LIVE IN A MIRROR, is the man doing the dance step: I, my grave his face, I: the woman with the wound at her throat, right and left in her hands the split bird, blood on the mouth, I: the bird who with the script of his beak shows

> the murderer the way into the night, I: the frozen storm. (Müller, 1989, p. 102)

Staging the subject of vision

Müller wrote *Bildbeschreibung* for the theatre. *Bildbeschreibung* is intended to be staged, but, it is only when one starts thinking about staging the text that the full complexity of the relationship in *Bildbeschreibung* between the one seeing and what is seen becomes evident, a true riddle.

According to the conventions of staging, it is necessary to make present on stage the world as given in the words on the page. Müller's text however, refuses such a notion. The labyrinthine structure, with its repetitive returns 'in difference', undermines the illusion of a transparent window giving access to a world 'behind'. Instead, the text draws attention to how words never quite coincide with what they refer to and how they always imply a deferral and consequently, difference. And also, how words provide a perspective on what they (try to) describe, and how descriptions involve focalization, or point of view.

It is not clear, moreover, who this speaking 'I' is, where he or she is, and how we are to understand the relation between what the 'I' is saying and the situation in which it appears. Except for the very last lines, the text is entirely lacking in personal pronouns and indeed, in these final lines, the 'I' appears only to disintegrate and disappear into the landscape. The text mimics the strategy, common in scientific writing, of erasing traces of subjectivity and the viewer's position in order to produce a 'view from nowhere' (see Fox Keller in Chapter 3). Yet, obviously, this is not an objective description. In this text, the absence of point of view draws attention to focalization rather than obscuring it. The text prevents automatic identification with a meta-subject that would provide a suggestive 'view from nowhere'. Instead, positioning is turned into a problem for the reader, and also into a problem for anyone staging this text.

Bildbeschreibung lacks any indication of character or situation. Is this text a monologue? Is it a description of what is to be seen on stage? The director must decide whether to stage the *Beschreibung*, or the *Bild*, or something in between. He or she must literally produce a vision of the text, deciding what the audience will see by taking a position and deciding on the relationships between words and images, the verbal and the visual. Staging this text means producing a reading of the description of the picture. The reading, in its turn, will imply a vision, a point of view, for the audience to take up. Seen in this way, *Bildbeschreibung* does not present an expression of the freedom of the spectator as a master of possible semiosis, but of staging as constructing a position for the audience as the subject of vision. The individual seer receives an invitation to take up this position, and, from it, to identify with the subjective vision presented. Whether the seer takes up the invitation – or not – will depend on the relationship between the desires, presuppositions and expectations that are part of the point of view presented to the seer, as well as on the way in which this individual seer as subject has been conditioned to see.

Wo ist der Mensch?

In 1997, Stefan Kunzmann presented a staging of *Bildbeschreibung* in one of the rooms of the *Hochschule für Angewandte Kunst* in Vienna.[3] It was his final examination project and marked the end of his education as a stage designer. The title of his installation, *Picture Description/Explosion of a Memory*, mirrors (albeit in reverse) the title of the English translation of Müller's text (*Explosion of a Memory/Description of a Picture*). The concept of mirroring was a constant in the installation.

While Kunzmann presented his 'staging' of *Bildbeschreibung* in a room of the art academy, the architectural features of the room, as well as the way he made use of these features, were reminiscent of a conventional theatre with a box-set stage. The room was rectangular and half of the floor was raised. The division between the levels of the floor was emphasized by a protrusion of the wall, reminding one of the proscenium arch in a traditional theatre. Visitors entered the room through a door that opened to the lower half. Looking towards the 'stage' they saw two video monitors, each playing a short loop of film. One showed a bird that appeared to be flying backwards and the other a barren landscape. On the wall behind the monitors was a text that was difficult to make out. As members of the audience came closer, they discovered that it was in fact a mirror image of a text. On the black, shiny surface of the wall in front of them, the audience saw the reflection of Müller's text. This was printed on the wall behind them, in white on black. Within the mirror image of the text, the spectator saw his or her own reflection and shadow projected onto the text. Like the videotapes, the lights in the room were programmed to perform a loop: they slowly brightened and then faded out in six-minute cycles. As a result, the mirror image of the text appeared and disappeared – the staging began and ended – in an ongoing repetition of the same. Figure 5 shows an impression of the installation, combining visions in two directions.

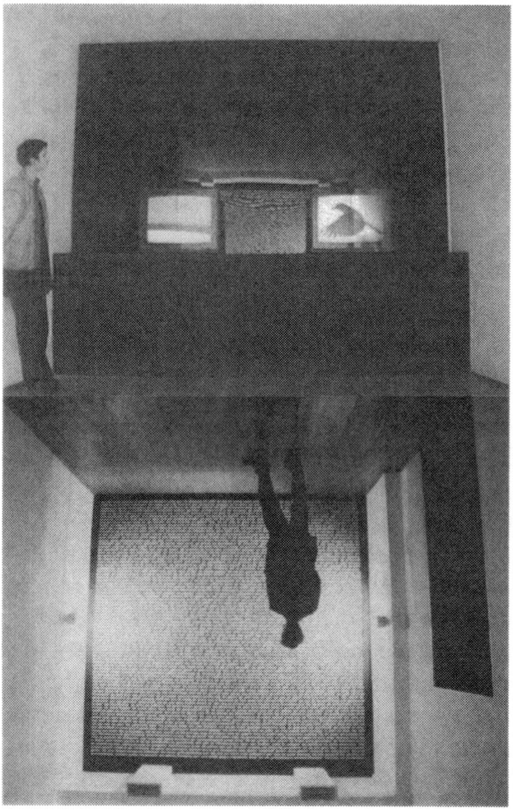

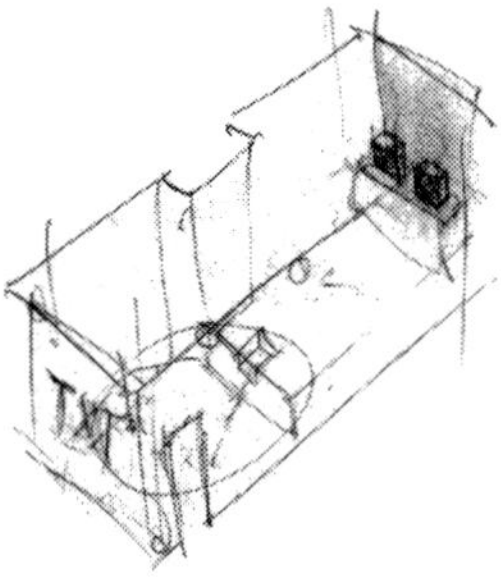

Figure 5 Impression by Stefan Kunzmann of his installation *Picture Description/ Explosion of a Memory*. Reproduced with permission of the artist.

'*Wo ist der Mensch?*' (Where is the actor? Or Where is the space for the actor?)[4] asked Kunzmann's supervisor, surprised that a student, graduating in theatre design, would present a piece of work that did not include a place for actors and was not intended to be used by actors. This was indeed surprising, since the presence of actors is usually understood to be one of the essential elements of theatre performance. Ordinarily, actors establish a relationship between the language of the play and the vision presented on stage, their embodied presence providing the audience with a visually marked position to 'step inside'. Stage design then, is a means of positioning the actor within a particular frame of reference. This may be done by means of illustrating and confirming the situation as given within a dramatic text, by referring loosely to the text, or even by contradicting or negating what is represented in the text. Kunzmann did none of these. Instead, he literally presented a staging of the text.

Caught between two mirrors

Kunzmann's strategy of making present the written text on stage brings to mind strategies used in the theatre of the 1970s and 1980s to undermine the positivism involved in theatrical representation through the infiltration of what has traditionally been the banished 'other' of dramatic performance: the written text. This was sometimes accomplished by having actors open books and read them on stage, through the use of written text as part of the setting, or through a performance of writing within the staging. In these examples, reading and writing are used as signs (or emblems) for the failure of 'presence' not only on the stage but also in the world where barriers 'always already' exist between text and meaning, intention and reception and between individuals. But in Kunzmann's installation, the making present of the text did not serve to show the audience that the real thing is to be found elsewhere, nor to contemplate its inevitable absence. Instead his installation redirected attention to the relationship between (both verbal and visual) text and a reader. In this installation the seer is a reader and this reader is at the same time the performer. With his or her performance, this performer performs the subject of the vision presented by the installation.

At first sight the 'performance' – the mirror image of the text – looked like a printed text, inviting one to read it, and that is exactly what most visitors tried to do. They tried to read the mirrored text by moving closer to it, but the closer they got and the more they tried to focus on the text in front of them, the more their reflections and shadows interfered. Deciding it was impossible to read the text, they adopted different viewing attitudes, including different positions in the room. They attempted to see the 'performance text' as a visual composition and tried to find a place where their shadow would not get in the way. The only solution was to avoid looking at the staging of the text altogether, and to turn around and face the text that was written on the wall behind the spectator. This text, printed in white on black, was quite readable. One was not troubled by one's own reflection or shadow. Yet here too, positioning turned out to be a problem, albeit in a different way. The text was printed on the wall in reverse, as if it was a mirror image. The text addressed the audience in such a way that it suggested they should turn around and look the other way. Caught between two 'mirrors,' the audience was compelled to choose its own viewing position.

In the installation-staging of *Bildbeschreibung*, the seer became the performer engaged in a process of (re)positioning him or herself in the space opened up between *Bild* and *Beschreibung*. In this way, Kunzmann managed to stage the tension created in Müller's text between the *Bild* and the *Beschreibung*, without falling back into the all-too-familiar word/image opposition. By showing text and performance as both visual and verbal he exposes the two as mutually intertwined in the (verbal and visual) behaviour of an audience. Furthermore, the spatial set-up of Kunzmann's staging of Müller's text exposed the relationships between Plato's parable of the cave as a 'vision machine' and conventional conceptions of staging texts, in which the performance is understood as a reflection of some originating 'presence', or meaning, given in the text. Kunzmann's 'performance' does not show some imaginary world given in the text, but instead literally a reflection of the text. His staging adds yet another layer to the description of a picture of a landscape, and it turns the description of the picture into an image of a description. At the same time, this 'performance' exposes someone else as well, someone who remains in the dark in Plato's parable, namely the seer. In doing so, his installation draws attention to the point of view from where the performance is seen and the relation between this point of view and what is seen. This point of view is exactly what remains the question in Müller's text.

Moving along

Bildbeschreibung consists of the description of an image. Indeed, the reading of the image is the only thing the reader of the text has to go on. The text gives no indication either of what is actually there to be seen, or of who is seeing it. Furthermore, the point of view from which the image is seen seems constantly to shift, making it difficult to imagine oneself in a particular viewing position in relation to the image. The only way to 'see' the landscape is to 'walk along' with the reading and take up the positions presented by it. This is further complicated by the structure of the text. Müller's five-page text is one long sentence, without full stops or capitals, which requires that readers also constantly relocate themselves in yet another way. How Kunzmann presented the text, widely spaced on the wall, reinforced the grammatical confusion of Müller's text. The words were so spread out that it was difficult to keep track of the linear narrative they represented and, as a result, the linearity of the text seemed to expand into a visual composition. With their reading behaviour, the spectators-as-actors performed precisely the choreography that Müller's text speaks of, namely the interaction between visual texts and individuals seeing and reading them, trying to make sense of them. The installation illuminated how this happens in relation to points of view presented by Müller's text and the performance, and taken up (or not) by spectators.

The behaviour of the audience in Kunzmann's installation reads as a physical demonstration of the seer's response to the address presented by the aesthetic logic of the post-dramatic theatre, in which unitary focus as it is given within dramatic theatre is deconstructed to give way to 'an opsis which is without hierarchical dependence' which results in a decrease in subjective control (Lehmann, 1997, p. 59).[5] The address presented by this new theatre leaves the spectator free to 'wander around' and see for themselves (instead of being firmly guided by the perspective provided by dramatic structure). Lehmann describes this aesthetic logic in terms of a choreo-graphical inscription in 'choral space': a 'space' defined by 'a multiplicity of voices, a "polylogue", a deconstruction of fixed meaning, a disobedience of the laws of unity and centered meaning' (Lehmann 1997, p. 57, and 1999, p. 262).[6] He proposes Gertrude Stein's notion of the landscape play to describe the ways in which this new theatre mediates in new relationships between seer and seen, and aims at evoking new kinds of perception.[7]

Elinor Fuchs, in her account of the development of twentieth-century theatre (1996), also invokes Stein's notion of the landscape play. She speaks of a 'weakening' of the subject on stage, the effect of which she terms 'the death of character', the 'de-authentication, the "absencing" in some sense of the speaking subject' (Fuchs, 1996, p. 72).[8] She observes that often performances have non-linear structures and are no longer concerned with individual characters or with a temporal progression, but rather with a total state or condition. Correspondingly, the spectator's focus on the stage is no longer convergent: it is darting or diffuse, noting some configurations, missing others, or absorbing it all in a heterogeneous gaze (Fuchs, 1996, p. 92).

In Kunzmann's installation, the seers become the actors performing a choreo-graphical inscription. His installation sets the seer free to 'walk the landscape stage' which, within this context, means literally to move through it, performing the choreo-graphical inscription mediated by the positions implied within the presentation of the text. Seen this way, the question 'Wo ist der Mensch?' touched the very core of Kunzmann's staging, be it in ways perhaps not anticipated by the dramaturge who asked the question.

His installation also serves as a commentary on the kind of subjective desires and presuppositions that guide the gaze of those seers set free to look for themselves. Setting the audience free to be masters of possible semiosis, the installation performed the promise presented by Plato's parable: that one day we will be made to see, which means to be freed from our fixed positions in the dark, gaping passively at a shadowy spectacle. Yet, freed from their fixed viewing positions in Plato's cave, spectators appeared not to be free at all from the desire for the kind of vision and ways of looking of which the cave is a metaphor. No longer tied to their fixed positions, the visitors of Kunzmann's installation found themselves looking for a position from which they could see without being confronted by their shadow. They found themselves looking for a place marked by absence, a viewpoint that did not implicate the seer in what was seen.

This longing is also represented in the text of *Bildbeschreibung*, by repeated references to the sun. The sun represents the point of view of the all-seeing eye, the detached observer, who can see everything as it is from afar, like the Cartesian disembodied subject. 'THE SUN is always there and TO ETERNITY' (Müller, 1989, p. 97). In metaphysical thinking from Plato to Descartes and beyond, the sun appears as the giver of natural light, that which, as Derrida observes, is understood as the source of 'the very opposition of appearing and disappearing ..., of day and night, of the visible and the invisible, of the present and the absent – all this was possible only under the sun' (Derrida quoted in Jay, 1993, p. 509). In his critique of vision in modern French thought, Martin Jay refers to Derrida's remarks on the metaphor of the sun in order to show how conceptions of vision and visibility are intertwined with notions of stability, sameness and circularity for which the daily appearance of the sun is a symbol (a circularity represented within Kunzmann's installation through the cyclical fading in and fading out of the light). However, as Derrida (and Jay) also point out, the sun metaphor is ambiguous, for the sun can be a source of illumination as well as the cause of blindness: blindness to difference, to otherness and to temporality as these are bracketed in the drive to become a viewpoint and nothing more.[9] Kunzmann's staging of *Bildbeschreibung* shows such blindness to be blindness to the self as the embodied locus of looking.

The stage as mirror

In 'Discovering the Spectator' Fischer-Lichte's calls *Bildbeschreibung* an example of the shattered mirror that postmodern theatre presents to its spectators: 'It consists of numerous disparate elements which, even as a whole, render no meaningful unit, can reveal no unifying image. The image reflected by post-modern theatre is one of many "Others"' (Fischer-Lichte, 1997, p. 58).

Fischer-Lichte reminds us of Lacan's account of subject formation through identification with the image of a body seen (in the so-called *mirror stage,* which will be discussed extensively in the chapters to follow). She points out that, long before Lacan, Helmuth Plessner formulated a similar discovery: 'one finds the Self through forming relationships with the Other. By viewing the Self in the mirror as other or even in the mirror of the Other, an image of self is composed' (Plessner, in Fischer Lichte, 1997, p. 58). She then continues, comparing the theatrical signs presented onstage to the mirror that the spectator reflects back as the image of the Other: 'In that the spectators, for their part, reflect this image, they enter into relation with their own selves' (Fischer-Lichte, 1997, p. 58). This mixing up of self and other is crucial to what Lacan describes as the mirror stage and produces the misrecognition that, in his (Lacan's) theory is the foundation of our sense of self. In Lacan's theory, however, we do not view the self in the mirror as Other, but instead (mis)recognize the mirrored image (i.e., the Other) for the self. This mis-recognition is crucial to the mechanism described by Lacan.

Kunzmann's installation confronts the seer with a reflection of themselves in their understanding of what they see. This image literally obstructs their reading of the performance text and, as their performance demonstrates, seeing their own mirror image is something the visitors would rather avoid. Instead, they opt for a position free from confrontation with an image of themselves. Thus, the spectators can lose themselves in the 'others' presented to them in the mirror that is the stage. Such forgetting of the difference between self and other requires a position like that of the scientific meta subject described by Fox Keller (see Chapter 3), turning the world into a theatre that can be (imaginarily) seen from a position marked by absence. It is this position that allows for the experience of the loss of self as the experience of a 'split subject, who simultaneously enjoys, through the text, the consistency of his selfhood and its collapse, its fall' (Barthes, 1977, p. 21).[10] The text of *Bildbeschreibung* testifies to precisely such an experience.

In *Bildbeschreibung* identification happens, not with the image of a body in a mirror, but with 'numerous disparate elements' of a landscape that 'can reveal no unifying image'. The 'I' speaking (mis)recognizes him- or her-self in the mirror presented to him or her by the picture of the landscape, and identifies simultaneously with various elements seen in this mirror. The result is an instance of absorption more radical than Diderot 'stepping inside' the picture of a landscape and forgetting that he is actually looking at a painting. Furthermore, this identification is not inspired by an image that presents the promise of a unity not yet experienced (as it is the case in Lacan's mirror stage), but rather results from the impossibility of getting the picture, that is, the failure to unify all elements of the landscape represented in relationship to a point of view. The text of *Bildbeschreibung* reads as an attempt to make sense of an artistic representation of a landscape (*Bild*) seen through interpretative procedures that aim at actively constructing the tableau. To follow Diderot (Chapter 2), the *Bild* does not present 'a whole contained under a single point of view, in which the parts work together to one end and form by their mutual correspondence a unity as real as that of the members of the body of an animal' (Diderot quoted in Barthes, 1977, p. 71). Instead, *Bildbeschreibung* testifies to the failure to bring all together, and this impossibility of getting the picture results in an instance of radical absorption. The 'I' speaking in *Bildbeschreibung* becomes the landscape, identifying with various elements simultaneously.[11] This result is not the jubilant moment of the Lacanian child, imaginatively 'becoming' the unitary body image, but is more like what Barthes describes as *jouissance*: a radically violent pleasure that evokes an experience of loss of self.

Aspiring to the landscape

Understood as a symptom that alerts us to the relationship between the visual address presented to a seer and the subjective desires, presuppositions and anxieties guiding the way of looking of this seer, the radical moment of absorption as described in *Bildbeschreibung* provides an understanding of the landscapes on the post-dramatic stage as appealing to the kind of longing that Petra Halkes terms 'aspiring to the landscape'. This is the desire for an absolute end to difference as it is embedded in the modern history of science and art, a desire that finds its uncanny articulation in landscape painting, that is, in pictures [*Bilder*] of landscapes. The history of landscape painting is enmeshed in a story of longing for seemingly opposite ends: nature's conquest and an affirmation of the self on the one hand, and on the other transcendence into nature and effacement of self. Both spring from a discontent with the inadequacy of human embodied existence and a desire to overcome its limitations.[12] Since the beginning of Western culture, this desire for the absolute has been framed by stories of Paradise and Arcadia, the image of the utopian ideal of the world as a garden in which we are one with nature. The fall from Eden signifies the beginning of a longing to return to a state of presumed original fullness. This meta-narrative has a double, its 'Other'. Paradise is framed within the nothing that God created it from, and the desert that remained outside of its gates. Hence, within the history of landscape, beneath the meta-narrative of 'Reinventing Eden', the desire to avoid representation and reach a primitive, undefined chaos outside of the garden forms another meta-narrative. The understanding that the garden is only a historical, time-bound consolation and that oneness cannot be found within the garden walls leads desire outside of the garden and into the frightful unknowable beyond (Halkes, 2001, p. 15–16).

Both sides of this notion of landscape can be seen at work in the use of landscape as a metaphor for the post-dramatic stage. Lehmann speaks of the 'longing for a space beyond telos', a space that is '"placed" on the borderline of logic and reason, on the threshold of what is thinkable and what is beyond reasoning', 'a pre-logical "space" that gives room to the play of being and becoming of all reality and precedes every possible distinction' (Lehmann, 1997, p. 56). Fuchs associates the landscape stage with 'the collapse of boundaries between human and world, inside and outside, foreground and background' (Fuchs, 1996, p. 93), and states that the postmodern artist longs for a vanishing natural world, or a vanished natural world that existed before history, before culture (Fuchs, 1996, p. 107). Here, Fuchs' post-modern artist longing for the 'natural world' seems to mark the idyllic end of the spectrum, while Lehmann's pre-logical space marks the undefined unknowable beyond.

In line with these observations, I propose to read Diderot's description of his encounter with the landscape paintings by Vernet, and Müller's *Bildbeschreibung*, as accounts marking two ends of the spectrum of aesthetic pleasures provided by the landscapes on the contemporary stage, each testifying to a different kind of subjective investment of the seer in the seen. A point of connection is that both involve a split subject, allowing for losses of control (e.g., Diderot's experience of not being controlled and therefore having freedom to wander and create his own stories at will, and *Bildbeschreibung*'s loss of self), while still allowing for this loss of control to be a subjective experience. In both cases, this results in a kind of internalized theatre in which the theatricality of the situation is precisely what needs to go unnoticed in order to present the seer with the sense of freedom.

For Diderot, his freedom to interact with various elements represented in the painting, and to make up stories about them, takes place within the vision of someone outside the painting, someone who is looking at a representation of a landscape but is allowed by that very representation to forget his or her bodily condition as a presence standing in front of a painting that represents a landscape. That is, his freedom to create meaning at will happens within a vision that includes all kinds of assumptions concerning artistic representations and what makes them convincing, and concerning possible meanings of what is encountered in them. It is within this vision that Diderot is free to wander around in the painting and give meaning (in the form of the stories) about what is happening in the landscape.

Seen this way, the relationship between Diderot looking at the painting and Diderot looking around in the landscape represented by the painting, illustrates the relationship between two different aspects of visuality, defined by psychoanalysis in terms of the distinction between the look and the gaze. The looking around of individuals takes place within the larger cultural-historical perspective called the gaze (which is discussed in depth in Chapter 7). Within this gaze we can seem to be free to look around as we wish, but this freedom is a result of our unawareness of the way the gaze guides what we see. This is not to say that Diderot's seeing of the painting is the gaze. Nobody 'owns' the gaze. Rather, the situation described by Diderot, in which he writes as if from a position within the painting, apparently free to go where he wants and give meaning at will, yet doing so within the vision of Diderot outside the painting, is in some ways analogous to the relationship between the gaze and the look.

Absorption here results from the way in which the painting makes Diderot forget about his position outside the painting through the presence of an address that corresponds to his wishes and demands (concerning the way landscapes should be represented, what is exciting and beautiful, how he wants to look at them and be taken along with what is there to be seen). As a result of this address he remains unaware of the perspective implied in his looking, and consequently of his own subjective position in relation to what he sees. Therefore, Diderot can 'step inside', feeling free to 'wander around', while at the same time occupying a position of all seeing seer.

Brought to bear on the landscape stage, Diderot's story points to the ways in which the freedom of spectator to absorb everything in a heterogeneous gaze results from an address that invites the seer to look around and see for him or herself, while providing the seer with a position that allows the seer to remain the blind spot in his or her own vision.

Bildbeschreibung, on the other hand, describes a situation in which a seer is confronted with the fact that freedom to see for yourself does not mean one is master of possible semiosis. The text testifies to the struggle of a seer to describe what he or she sees. Understood as the 'shattered mirror that post-modern theatre presents to its spectators', *Bildbeschreibung* points attention to the fact that the absence of coherent structure and unity may set the spectator free 'to associate everything with anything' or 'to refuse to attribute any meaning at all and simply experience the objects presented to them in their concrete being' (Fischer-Lichte, 1997, pp. 57–58). However, this does not mean spectators are free to see things as they want, or to or assign meaning at will. Although the non-unitary aesthetic logic of the landscapes on stage may suggest freedom to make up all kinds of stories in an attempt to get the picture, not all stories will do. The struggle of the 'I' in *Bildbeschreibung* to give an adequate description of what is there to be seen is reminiscent of Benveniste's critique of de Saussure's account of the arbitrariness of the linguistic sign. Benveniste points out that linguistic arbitrariness does not mean relativity, or that anything goes, which it has often been construed to mean. The relation between signifier and signified may be arbitrary if seen from the detached point of view of one limited to externally observing the bond established between an objective reality and human behaviour, or, as Benveniste evocatively puts it, 'under the impassive regard of Sirius' (Benveniste, 1971, p. 44) Seen from Sirius, the linguistic sign is arbitrary because the same animal is called *boeuf* in one country and *Ochs* elsewhere. But, as Benveniste points out, the real problem is far more profound. This real problem is that as soon we leave this position 'under the impassive regard of Sirius' the connection between signifier and signified is not arbitrary at all. From the perspective of an individual user of language the relation is necessary, and through their necessary connection the world appears to us 'as it is'. Similarly, although our condition as discursive beings implies that how we see what we see is relative rather than absolute, it does not mean that the effects of discourse are experienced as relative by those who participate in it. The discourse we are born into is not relative but formative for the world as it appears to us. And this discourse implies point of view.

Bildbeschreibung is about this limitation to our freedom and reads as a demonstration of the desire for the position of a meta-subject, a position from where the effects of confrontation with these limitations can be experienced as an aesthetic pleasure. In Kunzmann's staging, the mirror that is the performance confronts the audience with a reflection of themselves as blind spot in their reading of *Bildbeschreibung*. Only as long as we remain blind to the way this blind spot is involved in how we see what we see, can we have the illusion of being masters of possible semiosis.[13]

5
Navel Gazing as Critical Practice

'How can we know the dancer from the dance?' reads the famous last line of Yeats's poem *Among School Children*. This line has often been interpreted as a rhetorical question, which states the potential unity between sign and referent. It is however, as Paul de Man (1979, pp. 11–12) has pointed out, equally possible to read the last line literally rather than figuratively. Read in this way, Yeats's question is not just a confirmation of the fact that sign and referent fit together so exquisitely, but that all differences between them are effaced. Even more, 'How can we know the dancer from the dance?' becomes a serious question that sets out to deconstruct precisely the intricate intertwining of sign and referent in the 'presence' that the poem addresses, and thus to expose the conflation that allows for the rhetorical reading in the first place. In this second reading, the question is not whether or not we can know the dancer from the dance, but how we can know the difference. The final line is read literally as meaning that, since dancer and dance are *not* the same, it might be useful, perhaps even desperately necessary, to tell them apart. Furthermore, the two possible readings of the poem do not simply exist side by side. They have to engage each other in direct confrontation, for the one reading is precisely the error denounced by the other, and has to be undone by it. Therefore, de Man calls the question 'How can we know the dancer from the dance?' a figure that cries out for the differentiation that it itself conceals (de Man, 1979, p. 12).

First, from this perspective, the question 'How can we know the dancer from the dance?' touches a point most relevant with respect to the contemporary stage, where bodies and other signs seem to 'break free' from the dramatic frame in order to present 'themselves' instead of functioning as a sign for something else, often suggesting an inseparability of sign and referent similar to that suggested by the rhetorical reading of Yeats's poem. This results in a paradoxical situation: On the one hand, accounts of new developments in the theatre testify to a growing awareness of the inevitable discursivity of our ways of seeing and understand what we see as being the product of signifying practices rather than simply that which is 'there to be seen'. On the other hand, there is a persistent tendency to understand the effects of this deconstruction in terms of increased authenticity, immediateness (the thing brought back to itself) which is in many ways analogous to the inseparability of dancer and dance in the rhetorical reading of Yeats's poem.

Second, the question 'How can we know the dancer from the dance?' seems to be particularly apt within the context of developments in the theatre in the past decades. In this theatre of the recent past, dance and dancers have presented a strong appeal to the desire involved in the rhetorical reading of Yeats's question; that is, the desire to come to a point where performers are 'themselves' in events in which signs and meaning are no longer separable. Yeats's question brings to mind many examples of theatre makers who began to collaborate with dancers and choreographers, or to experiment with making dances themselves, as well as a more general blurring of boundaries between theatre and dance. In this theatre, Yeats's question 'How can we know the dancer from the dance?' rings out with urgency precisely because of the duplicity observed by de Man. Although 'we' today seem to know well enough that there is no sign without referent, it is very often precisely the apparent conflation of sign and referent that makes this theatre so appealing. Therefore, in order to understand the relation between seer and seen as it takes place in this theatre, neither does it suffice to observe that dancer and dance, actor and act, sign and meaning seem to be conflated in an undivided 'presence', nor does it suffice to deconstruct this 'presence' in order to undo the apparent conflation. What is needed instead is to understand the relationship between seer and seen in terms of the duplicity observed by de Man. Understood this way moments of apparent conflation can turn into meaningful pointers, bringing attention to the subjective point of view in what seems to be 'just looking' at what is 'there to be seen'.

In this chapter, I approach this subjective point of view through a discussion of Roland Barthes's famous text *Camera Lucida*, and *Bas and Elze Dance* (1996), a theatre performance that presents a self-reflexive metatheatrical commentary on the developments in the theatre described above. This discussion leads me to propose the concept of the navel as a critical tool in reconnecting moments of apparent conflation of sign and referent with a subjective point of view. The anatomical navel is a scar, a leftover that is itself meaningless, but carries meaning through its ability to focus attention on the initially vital connection that was cut in order for the human being to become an independent entity. Similarly, the navel as critical concept can be an index pointing to a lost relationship, cut through in order to produce the subject as a disembodied eye/I that is 'just looking'. Used as such an index, the navel can help reconnect what is seen to a historically, culturally and socially positioned seer – a viewer, furthermore, who is a body and therefore marked by mortality.

The Bas Raadsheer versus Elze Struijs case

In *Bas and Elze Dance* actors Cas Enklaar and Els Ingeborg Smits play two fictive fossils of theatre history. The title of the show refers to an earlier production titled *Bas and Elze Look Back* (1992), in which the characters Bas and Elze looked back upon their careers.[1] They have worked with all of the famous members of the theatre community and have played most famous, and infamous, plays. Although their bodies threaten to fail, they still know all the lines by heart. Both live in the Louise Houbée Foundation, a (fictitious) home for elderly artists. When we meet them again in *Bas en Elze Dansen*, they are working on a new production of Sophocles' *Electra*. Elze has done *Electra* before, back in 1936, for which she received a prestigious theatre award. But now she does not feel satisfied by repeating Sophocles' words. This time, she wants to dance *Electra*. She quotes Martha Graham's famous dictum that 'the body does not lie' and dances *Electra* with a can of coffee beans on her head instead of an urn.

Bas is shocked; Sophocles' words have proved their value for over two thousand years. What would we know about *Electra* if the words had not been preserved and handed down to us? Maybe the body does not lie, but very often it is hard to see what it says, and it also has a rather short life. Moreover, given Elze's age, it is not very likely that we will be able to enjoy her truthful bodily message much longer. Whereas Elze's body will die, the words of Sophocles will be handed on to coming generations, to be performed by ever-new bodies.[2]

If it were up to Bas, Elze would stop moving about immediately. Movement compromises her delivery. While Elze does some warming up exercises, Bas memorizes a list of 43 rules for the actor as recorded by Cor van der Lugt Melsert in 1949.[3] These rules are all about self-control and control of bodily appearance. One rule reads that the actor, unlike the musician or the sculptor, has only himself at his disposal: his voice, his body and his presence. On stage, the actor has to do without a musical instrument or a block of marble to hold on to. This being confined to his own body seems to be hard for Bas to accept. He identifies with the undying words of immortal drama-authors and hides his own corporeality, shielding it behind an obsessive preoccupation with manners and looks. For him, what counts in the theatre is not his own physical presence but the words of the dramatist. His body is to be controlled and, metaphorically speaking, kept out of sight, in order to prevent it from distracting the audience's attention. His physical presence has to be made subservient to the architecture of the drama that frames it and makes it readable in terms of the telos as given within the text. Elze, on the other hand, immediately begins to peel off the layers of gilding to show, *sans gêne,* an elderly body dressed in a rehearsal outfit that does not exactly become her. The exercises that she performs, instead of showing off physical prowess, bear witness to failure and aging and to what she can no longer do. According to her, it is this body, this carcass as she calls it, that speaks more truly than Sophocles' words will ever do.

The rules for the actor as memorized by Bas reveal the kind of technical training that these fictive historical actors (supposedly) received early in the twentieth century. They say a great deal about the changes in the conception of acting since then. As dramaturge Marianne van Kerkhoven (1998) puts it

> It is no longer the technical ability to put oneself in someone else's position, but the personality of the actor as human being that is placed in the forefront. This trend towards disappearance of the character is further reinforced by influences from dance and performance art. (Van Kerkhoven, 1998, p. 111; my translation)[4]

Van Kerkhoven defines the new actor born out of these developments in terms of a 'third' form of acting:

> [I]n contrast to 'the Stanislavski actor' whose work is based on immersion in the character and 'the Brecht actor' who displays his character to the audience, the 'third variation actor' primarily wants to show himself to the audience, whether or not *by means* of a character. (Van Kerkhoven, 1994, p. 10; italics in the original)

In this development, the work of the New York based Wooster Group is often mentioned as important influence. Van Kerkhoven, also, refers to the Wooster Group, quoting actor Ron Vawter's description of acting as 'finding out who I am in front of an audience' (Van Kerkhoven, 1994, p. 12). The actor no longer steps into the character's shoes, but rather allows this imaginary figure to live within himself; the character almost 'disappears'. It is worthy of note that Ron Vawter, in search of himself in front of an audience, usually spoke the words of others. He spoke the words of Roy Cohn and Jack Smith, for example, as well as those of many fictional characters. Rather than presenting himself in his own words, his search for who he is problematizes the very notion of self and shows it to be inseparable from the discourse in which it is expressed. This complication is, as pointed out above, a recurring theme in *Bas and Elze Dance*.

Bas and Elze Dance foregrounds the relationship between the physical self and the words of others as an important motive behind the search for new modes of theatrical expression. The performance demonstrates that while theatre may be as old as Methuselah, it is, nevertheless, alive and kicking. The characters Bas and Elze, with all their memories of a theatre long gone, represent theatre history, just as the would-be rehearsals for their new show represent the continuous search for new modes of theatrical expression. The questions stated in the programme leaflet: 'Is dance to be preferred to the spoken word? Does the body speak more truly than language?' recall the move towards dance so typical of many experimental theatre performances during the last decades of the twentieth century.

Much more, the performance presents a sampling of strategies of de-theatricalization as they have proved successful in the deconstruction of dramatic theatre. What we see on stage is not an illusionist representation of another world, but the empty space of a small black box theatre: Peter Brook's empty space.[5] Instead of a stage set, a few props are scattered around: a table, some chairs and a coffee machine. An old painted backdrop showing a winter landscape hangs, somewhat out of place, against the back wall. Two actors dressed casually enter the stage and address the audience. They introduce themselves as actors and explain what they are going to do. This is not going to be a polished performance. Instead, they are going to show us a rehearsal where everything happens *à l'improviste*. The actors show fragments of the play they are rehearsing, try out things with the text, discuss their roles and the interpretation of the play, disagree, make coffee and smoke cigarettes. They are constantly there as 'themselves', commenting on the roles they play, or using elements of the characters to reveal things about themselves.

However, instead of using these strategies to de-theatricalize what we see on stage, this performance actually stages such strategies, thereby retheatricalizing what we think we see. The actors seen as 'themselves' on stage are not Cas Enklaar and Els Ingeborg Smits, but Bas and Elze. The actors 'themselves' are characters. The performance demonstrates how the deconstruction of the dramatic theatre serves as framework that produces the actor's presence as him or her 'self', and how seeing this self depends upon a frame of reference. The performance also shows that deconstructing or 'breaking through' such a framework does not open onto some undivided presence 'behind' it. On the contrary, the frame is shown to be constitutive of this very distinction. *Bas and Elze Dance* demonstrates that the multiplication of the frame does not result in increased visibility of the thing in itself, but rather in indeterminacy. Furthermore, this indeterminacy undermines the position of the seer as the one who is able to see it 'as it is', and helps to expose the influence of framing in what we think we see.[6]

Punctum

On stage, we witness a day of the rehearsal process. Bas and Elze rehearse, discuss, have lunch and continually reminisce. They are more than a hundred-years old now and still remarkably energetic. Nevertheless, every now and then age begins to show. They are short of breath, they fall asleep spontaneously and Elze sits down next to a chair instead of on it. They make mistakes and repeat themselves. Such 'involuntary' behavior on stage attracts attention, similar to what Barthes (in *Camera Lucida*) describes as the *punctum* in photography. Punctum denotes a moment of apparent conflation of sign and referent. The punctum is a detail that sticks out from the frame of symbolic reality. It describes a moment that seems to escape coding and intention, thus evading representation.[7] Barthes opposes the punctum to *studium*, which he describes as 'the extension of a field which I perceive quite familiarly as a consequence of my knowledge, my culture: this field can be more or less stylized, more or less successful, depending on the photographer's skill or luck, but it always refers to a classical body of information' (Barthes, 1993, pp. 25–26). Studium evokes a kind of general interest, but not the emotional stirring caused by the punctum, because in the case of the studium, emotion 'requires the rational intermediary of an ethical and political culture', it results from 'an *average* affect, almost from a certain training' (Barthes, 1993, p. 26).

The behaviour of Bas and Elze contains many symptoms of involuntary behaviour that seem to 'break through' the frame of the theatrical representation and fall outside the symbolic reality on stage. However, in *Bas and Elze Dance*, these moments are staged. These 'symptoms of involuntary behavior' that seem to 'break through' the symbolic reality represented to show a trace of the real presence of the actor 'behind' it are not traces of Cas Enklaar and Els Ingeborg Smits' real selves. Rather, they function as signs of unintended behavior of the fictional characters/ actors Bas and Elze. As such they do not rupture the studium but support it, except in one instance.

At a certain point, Bas and Elze repeat an entire scene we have seen shortly before. It is a scene in which Bas and Elze recollect memories, or stories that have to pass for spontaneous memories. There is something odd about this repetition. In this scene, they remember a hilarious moment that happened in a provincial theatre where all the actors began laughing so hard that the performance had to be cancelled. The scene starts from an embrace. They find comfort in each others arms and then both begin to laugh, saying: 'yes, I know what you are thinking'. They start telling the story together, finishing each other's words, and having a lot of fun in telling it. After this spontaneous memory, others follow and then, unexpectedly, they repeat the entire scene about the performance that was cancelled in exactly the same way as we have just seen it.

Now, if the repetition of this scene is supposed to be a sign of these two old people's failing memories – that is, if this is supposed to signify that the characters Bas and Elze mistakenly tell the same story twice – then it is strange that their memories fail in exactly the same way at precisely the same moment. Furthermore, it seems strange that their telling of this story, which is supposed to represent a spontaneous interaction between the two, is repeated in exactly the same way as we saw it shortly before. Performed the way it is, it looks more like the real actors Cas Enklaar and Els Ingeborg Smits have forgotten that they have played this scene already and mistakenly perform it a second time, exactly the way they have rehearsed it. The embrace from which the scene starts is a recurring motive in the play, which would make it understandable that they, by mistake, start the wrong scene.

But then, the iron discipline with which they do the whole scene again exactly the same way, with the same timing, movements, hesitations and slips of the tongue, is somehow at odds with the atmosphere of spontaneity suggested by the performance. Moreover, why would two people make the same mistake at exactly the same moment? If one of them indeed inadvertently started to repeat a scene, and the other decided to go along with it in order to keep the show going, it would make more sense to do a variation on the scene in order to give it an improvized character similar to the rest of the performance.

The repetition does not fit within the logic of the story of Bas and Elze. Equally problematic, it appears to understand this repetition as a trace of the real presence of actors Cas Enklaar and Els Ingeborg Smits behind the mask. Precisely this undecidability, this impossibility to fit it into either side of the binary opposition that still defines much understanding of theatrical meaning production, makes it useful for a deconstruction of both.

The strictly studied character of the repetition not only highlights the sign aspect of the spontaneity, and the 'just being there' ambiance of the rest of the performance. It also confronts the seer in the audience with the 'frames' at work, that is, with his or her expectations, knowledge and desires. This instability causes a short circuit between actors and characters, thereby multiplying the frames. What actually do we see here, the actor or the character? This critical move undermines the opposition of framed versus non-framed, of symbolic representation versus real presence. It undermines the idea that breaking up the frame will result in a non-framed situation, opening onto some real, previously obscured presence. Instead, it leaves the audience in uncertainty about how to look, how to understand what is presented. It makes the audience aware of its own visual habits as they are involved in seeing theatre performance. In doing so, the repetition functions as what Mieke Bal has termed the navel.[8]

The navel

Like Barthes's punctum, Bal's concept of the navel refers to a tiny detail, a pointless point that somehow falls outside the logic of the scene represented. This does not however, bring Bal to a reading of such details as traces of the real. Instead she takes its appearance as 'just there to be seen,' as a meaningful symptom that draws attention to the limitations of signification, including the role played by subjectivity.

Initially, such a detail appears as noise within the represented scene. However, as Bal demonstrates, precisely because it is noisy, a detail can function as a meaningful pointer that helps to shift attention from the story represented to the story involved in the act of seeing this representation, the story of the interaction between seer and seen. Therefore, the navel presents a starting point for a re-narrativization of 'just looking,' by showing how 'just looking' involves a story emanating from a subject. Since each seer brings his or her cultural baggage, there can be no such thing as fixed, predetermined or unified meaning. This does not mean however, that there are no limits. All too easily, increased awareness of the subjectivity involved in what we think we see gets associated with relativity, with anything goes, and with increased freedom to give meaning at will. In fact, notions of the subject and subjectivity call into question precisely the autonomy, freedom and stability that define the position of the individual.

Effects of this misunderstanding can be seen at work in the use of Derrida's concept of dissemination to deconstruct fixed meanings and objectively given truths. Dissemination has proved to be a powerful tool to undermine persistent tendencies to explain what is seen in terms of origins, and for redirecting attention to signs as events that take place in interactive situations with someone for whom they appear as signs. Dissemination, however, does not mean unaccountability or the freedom to give meaning at will. Bal points out that Derrida himself has argued against this misunderstanding. However, as she also points out, this misunderstanding is somehow already implicated within Derrida's use of metaphor.

> Although [Derrida] undermines the phallic view of sign and meaning inscribed in Saussure's semiotic, Derrida is also implicated in it. This is because his dissemination, intended to dissolve the penetrating power of the dualistic sign, sometimes looks like an overwhelming dispersion of semen: coming all over the text, it spreads out so pervasively, so biblically, that it becomes like the stars in heaven or sand at the seashore: a promise to global fatherhood [...]. (Bal, 2001, p. 82)

It is this global fatherhood that Bal argues against with her notion of the navel as a centre without meaning.

Bal introduces her concept of the navel in her book on Rembrandt's paintings – which mostly represent biblical stories. Bal demonstrates how these stories function as a frame in a manner analogous to the logocentric architecture of dramatic theatre. The story presents a frame that sets out a relationship between seer and seen, a frame through which all elements seen are given clear meanings, and can be understood in a meaningful relationship to one another. The frame is a guide to interpreting what is seen as a function of the story. Bal demonstrates this type of reading with a discussion of Rembrandt's *Danae* (1636; see Bal, 1991, pp. 18–24). The painting represents a pre-given story; various elements of the painting offer a visual representation of various elements of the pre-text. This is the story of Danae, barred from love by her frightened father but, in the moment the picture presents, visited by Zeus in the guise of a shower of gold.

Bas demonstrates that a reading of the same painting as a visual text, as an address oriented to a viewer as subject of vision, tells a different story.

> Looking at the picture, we see a female body, nude, displayed for the lust of the viewer who is allowed to peep into the intimacy of the doubly closed bedroom. But this is a visual story, the story of vision in Western culture. It is the story of the male voyeur and the female object, of the eroticisation of vision: it is the story of the central syntagm–subject-function-object–in which the positions are fixed along gender lines, through which, indeed, gender itself is constructed. (Bal, 1991, pp. 19–20)

The painting as a visual composition presents an address to a viewer, and this address interpellates the viewer into a position as seer in relation to what is seen. This other story does not deny or replace the story of the pre-text, but it does present a perspective on various elements of the image and the way things are shown. It also helps us to understand why this particular visualization of the story of Danae is convincing, and to whom. The fact that the painting representing the story of Danae addresses the viewer as voyeur looking at a naked female body, displayed for his visual pleasure, does not mean that the story of Danae is about voyeurism, nor that the visual story presented by this painting can be reduced to voyeurism. Rather, becoming aware of how the story represented here (the pre-text about Danae and Zeus) and the address presented to a viewer by means of this particular visualization are related, provides insight into how the meaning of this painting comes into being as an effect of the interaction between seer and seen as mediated by the perspective presented by the painting.

Bal demonstrates how in this process, a tiny detail that seems to fall outside the logic of the representation or appears simply as noise within it, can shift attention from a reading of visual signs in terms of a relation to a (supposed) origin (the story of Danae in Bal's example, or the drama text, or the self of the actor on stage) towards a reading of visual signs as events taking place in a historically and socially specific situation, and in relation to a seer. Such a detail can alert us to how the painting is organized as a visual text in respect to a seers, as well as to how what appears to be convincing must be understood in relation to the parameters that define seers as subjects seeing.[9]

The navel as the insignificant detail or noise is meaningful in two interrelated ways. First, visual 'noise' says something about the subject seeing because it is that which must be repressed in order to achieve the effect of unproblematic 'just looking'. Visual 'noise' has to be repressed in order to confirm the seer as the one who is able to see it 'as it is'. Second, such a detail – provided that it is not understood (and repressed) as noise but taken as a sign that addresses a seer – can be useful in relating what is 'just there to be seen', to the point of view from where it appears, thus exposing the point of view that gets obscured when these signs are repressed as noise. Seen this way, the navel functions as vision shifter, elucidating the relationship between the seer 'just looking' and what is 'there to be seen' in terms of the duplicity observed by de Man.

The navel of the dream

Bal is not the first one to have invoked the navel as an interpretative metaphor. Freud uses the term 'the navel of the dream' to designate a moment that seems to resist interpretation and that must remain obscure. He introduces the term in an analysis of one of his own dreams, a dream about a female patient named Irma. His psychoanalytical treatment relieved her hysterical anxiety but not all of her somatic symptoms. Unwilling to accept the solution Freud proposed, she breaks off the treatment.

In his dream, Freud takes Irma to the window and looks down her throat to see a big white patch and whitish grey scabs covering a wound. In his explanation, Freud interprets the figure of Irma as a condensation of various women he knows. He understands the patch and the scabs as a composite index for his anxiety about his eldest daughter's serious illness and his own ill health due to the use of cocaine. However, in a footnote to his interpretation of his own dream, Freud states that

> I had a feeling that the interpretation of this part of the dream was not carried far enough to make it possible to follow the whole of its concealed meaning. If I had pursued my comparison between the three women, it would have taken me far afield – there is at least one spot [Stelle] in every dream at which it is unplumbable [*unergründlich*] – a navel, as it were, that is its point of contact with the unknown [*mit dem Unerkannten zusammenhängt*]. (Freud, quoted in Bronfen, 1994, p. 82).

Elisabeth Bronfen (1994) makes an analogy between Freud's concept of the navel of the dream and the vanishing point in perspective. The vanishing point marks a limit we cannot see beyond, the point where receding lines meet. It marks the limit of what can be seen while, at the same time, it signifies an inaccessible beyond. The vanishing point mirrors the vantage point, the point from where the scene is depicted. It has its function within a logic that serves to support a position for the viewer as the one who can see it 'as it is'.

Bronfen explains how, in Freud's text, the notion of the navel of the dream as vanishing point serves to assuage the blow to his analytical potency constituted by Irma's cancelling treatment. She argues that the double navel (the metaphor 'the navel of the dream' and the representation of the white scabs) mediates two moments in which Freud's dream articulates a wounding of his sense of potency, by confronting him with a moment that eludes it. Irma's resistance 'castrates' him in the sense that it points out his impotence as an analyst; it marks the moment of failure in his interpretative system. In his dream, this wound manifests itself in the white scabs as an index for his anxiety about illness. In his analysis of his own dream, the fear for his own impotence manifests itself in his explanation of the white scabs as the 'navel of the dream,' a rhetorical gesture that relieves him of the responsibility to understand what he sees as a sign, and to know how to interpret this sign.

In the interpretation of his own dream, Freud constructs a narrative explicitly acknowledging that his interpretation was not carried far enough because he insists that the concealed meaning would take one too far afield. He justifies his interpretative impotence by stating that this 'tangle of dream thoughts cannot be unraveled' and that a reading of it 'adds nothing to our knowledge of the content of the dream.' He contents himself instead with the figure of 'a spot where the dream recedes down into the unknown' (Bronfen, 1994, p. 83).

> At stake then when we encounter the navel of a representation – the knotted scar covering and recalling a wound – lies the issue of impotence and the desire to reassert potency; resistance of an enigma and its encroachment upon the subject; failure and exculpation; castration as it structures symbolic relations, and sublimation as it diffuses the threat of the Real. (Bronfen, 1994, p. 85)

This sublimation tames the real, turning the lack of signifier (where words fail) into a signifier for lack ('navel of the dream'). Significantly, this threatening lack of signifier, which has to be sublimated and made harmless, is marked in Freud's dream by the whitish grey spots that show themselves to Freud as he 'penetrates Irma's oral cavity, leading him down into the unknown of the body' (Bronfen, 1994, p. 83). This relationship between the vanishing point, female bodies, and the need to sublimate a lack of signifiers into signifiers of lack will be explored in the next chapter, where the analogy between the navel as the elusive spot, and the vanishing point as point of origin in perspectival vision, will help to expose the body involved in 'just looking.'

By acknowledging the navel of his dream as the vanishing point, as the place where interpretation would lead too far afield, Freud restores his position as viewer with the potency to distinguish between what is meaningful and what is not. The potency to make this distinction is what is at stake in *Bas and Elze Dance* as well. The narrative moment, which I have called the navel of the performance, presents the viewer with a confusing knot that undermines the binary opposition of representation and presence and makes this opposition felt as part of the logic that structures what we think we see. This moment resists an automatic reading of what is seen in terms of the story represented. It also resists a reading as a moment of exposure of the real presence of the actors as 'themselves' behind the representation. This moment is confusing not because the frame is taken away and the thing appears in all its glorious presence. Rather, confusion results from the way the performance prevents 'just looking' from taking place, and leaves the viewer in confusion about what he or she is seeing.

Like the navel of Freud's dream, the navel of this performance is a small detail that can easily be suppressed as meaningless, as a moment that would lead one astray if one were to try to fully understand it. Actually, ignoring this moment smoothens the logic of the story presented. The story makes complete sense and is much more coherent without it. Within this coherent story, it is noise, an irritating moment that seems to ask to be covered up like a mistake or a stain. Understood as the navel of the performance, however, it can function as an index by drawing attention to the relationship that has to be suppressed and sublimated, in order to produce the seer as independent subject 'just looking' at what is 'there to be seen'.

Always already dead

Barthes wrote *Camera Lucida* while preoccupied with 'an "ontological" desire' to know what photography is 'in itself' (Barthes, 1993, p. 3). He finds this to be an essential feature in what for him is the quintessential photograph: an image of his mother. He describes how, after his mother's death, he looks through her photographs 'looking for the truth of the face [he] had loved' (Barthes, 1993, p. 67). Finally, he finds what he is looking for in a photograph of his mother as a child of five, in a winter garden with her brother. In this photograph he finally finds a trace that exposes his mother as she really was; that reveals part of her essence or self. Barthes does not understand this quality in terms of information about his mother given by the photograph or by the way the photograph represents her. Rather, he understands the impact made by the photograph in terms of its ability to go beyond what he termed the studium and present him with what Lacan calls the Touché, the Occasion, the Encounter, the Real, in its indefatigable expression (Barthes, 1993, p. 4). 'In a first impulse, I exclaimed: "there she is! She's really there!"' (Barthes, 1993, p. 99). The photograph then, seems to appeal to a desire for immediateness, directness, a desire to go beyond the 'always already' of the symbolic order.

> [F]inally the Winter Garden Photograph, in which I do much more than recognize her (clumsy word): in which I discover her: a sudden awakening, outside of 'likeness', a *satori* in which words fail, the rare, perhaps unique evidence of the 'So, yes, so much and no more'. (Barthes, 1993, p. 109; italics in the original).

'Look', 'See', 'Here she is', the photograph seems to say, a gesture Barthes associates with the gesture of the child pointing his finger at something and saying: that, there it is, lo (Barthes, 1993, p. 5).

Bronfen reads Barthes's account of the punctum as a specular counterpart to Freud's navel of the dream. Like the navel of Freud's dream, the punctum points to what evades representation. It is disturbing precisely because it cannot be named. It is a 'blind field' (Barthes, 1993, p. 57). But unlike Freud, who marks the navel of his dream as that which must be left obscure, Barthes recognizes the punctum as the spot, which precisely does not lead astray. Barthes writes:

> I exhaust myself realizing that 'this has been.' ... I passed beyond the unreality of the thing represented, I entered crazily into the spectacle, into the image, taking into my arms what is dead, what is going to die. (Barthes, 1993, pp. 107, 117)

Barthes understands the punctum in terms of a facticity that allows him to go beyond representation, the effect of which he describes as a being absorbed into what he sees ('I entered crazily into the spectacle'), to be united with that which the spectacle presents to him. Barthes enters crazily into the spectacle to be united with what is both already dead (his mother), and what is going to die (the five-year-old child in the photograph). The photograph shows him the dead in a way that can momentarily forestall death, allowing him to embrace these always already dead people as though they were living. In this respect, he creates a relationship between photography and theatre as a means of making present here and now what is always already dead.

> We know the original theatre and the cult of the Dead: the first actors separated themselves from the community by playing the role of the Dead: to make oneself up was to designate oneself as a body simultaneously living and dead: the whitened bust of the totemic theatre, the man with the painted face in the Chinese theatre, the rice-paste makeup of the Indian Katha-Kali, the Japanese No mask. [...] Now it is the same relation which I find in the Photograph; how 'lifelike' we strive to make it (and this frenzy to be lifelike can only be our mythic denial of an apprehension of death), Photography is a kind of primitive theatre, a kind of *Tableau Vivant*, a figuration of the motionless and made-up face beneath which we see the dead. (Barthes, 1993, pp. 31–32; italics in the original)

Photography has its function in warding off death, postponing it. Photography can make present the dead while at the same time photographic representation brings out death in what is living by turning the living into an image that represents the living as always already dead. The photograph is a representation of a moment that per definition is no more, is absent; it shows what is already dead. This character of always already being dead has to be countered by, suppressed by, or obscured through strategies of life-likeness that obscure the difference between sign and referent. If successfully obscured, this results in a moment that stands out like a tableau and is no longer experienced as representation, but as presence here and now. In these photographs, sign and referent seem to coincide and as a result, the distance between seer and seen is likewise transcended.

What if we take these always already dead people embraced by Barthes as though they were alive – an illusion afforded by the particular characteristic of the photograph that allows him to 'enter like crazy' – as symptomatic of a more general condition summarized in Derrida's dictum 'always already'? Then, the photographic punctum describes the moment at which we see what we know to be always already representation, and therefore absent, as 'just there to be seen', lo, this here, present. In this moment, the always already dead are experienced as being present, a condition that is evoked by the punctum as a moment where sign and referent seem to become conflated. This moment allows Barthes to 'enter like crazy', it draws him in, absorbs him into what he sees.

The photograph, or more specifically, photographs characterized by punctum, appear as a fetish: we know very well but ... . We know very well that what we see is always already dead, that it is a representation of something absent, yet some photographs allow us momentarily to enter this absent world, to be absorbed into it and experience what is seen as present, as here and now, and as the thing in itself rather than as a representation. Similarly, the theatre can function as a fetish. Even though we are aware of the always already dead character of what we see, of its dependency on what is not there, the theatre nevertheless presents us with momentary experiences of presence, of immediateness, that seem to escape the realm of the always already constructed.

Re-animation

In *Bas and Elze Dance*, theatre likewise appears as a means of warding off death. The theatre appears as a means to make present and keep alive what is dead and/or absent: the words of immortal dramatic authors. Bas and Elze's urge to keep on making theatre seems to be a way to distract attention from death's approach, maybe even a means to keep death at bay. As soon as they stop rehearsing, death urges itself upon them and they begin to contemplate old age, their past and mortality.

Elze proposes dance as an alternative to the always already dead words of dramatists. She wants to make Electra present in a way more directly touching than Sophocles' words. She favours dance as a means of presenting a more intense address to a viewer, more intense than a representation of the text. Although at first Bas does not seem taken with the idea of dancing instead of talking, in the end they do dance. On several occasions during the performance, they perform dancing movements together, and the show ends with a dance sequence of several minutes.

These dance sequences are not motivated by the story and they seem to happen 'spontaneously'. During these sequences, the actors Cas Enklaar and Els Ingeborg Smits no longer perform the signs of age that characterize their impersonation of Bas and Elze, again resulting in ambiguous effects. What are we seeing here? Are we supposed to see Bas and Elze who have miraculously transcended the ailments of old age? Or do we see Els Ingeborg Smits and Cas Enklaar as 'themselves' performing these movements? Do we see actors who have freed themselves from the restrictions imposed upon them by the story or do we see characters who, by means of dance, transcend the limitations imposed upon them by the 'reality' of their life, old age and mortality? Again, these moments remain ambiguous. But in any case, dance appears to be a strategy that holds the promise of breaking free from symbolic reality, be it either the reality represented or the reality of representation. The performance is a practical demonstration of how dance actually manages to produce the effect of heightened directness, and is thus able to achieve an effect that theatrical representation cannot produce. Dance appears as a means of leaving behind Barthes's studium, and to reach out towards the audience in a way similar to the punctum.

According to Barthes, the punctum is the moment that does not lead astray because it marks the moment that counts, the moment that seems to be the fulfillment of a desire. In Barthes's text it is represented by his desire to retrieve a connection with what is lost: his mother. The punctum marks a moment that contains the promise to overcome the separation of which the navel is the scar: the cutting of the umbilical chord that produces the individual by separating him or her from the mother. The punctum thus appears, to cite de Man, as 'a figure that cries out for the differentiation that it itself conceals'. The navel, on the other hand, helps to understand the punctum in terms of the duality de Man describes, namely, as the effect of signs that derive their meaning from something absent. This status, however, remains concealed precisely by the address these signs present to a particular viewer.

The relationship with the seer is what has to remain concealed in order to produce the effect of the punctum. This relationship is also what remains concealed in Barthes's text. The punctum is 'what I add to the photograph and what is nonetheless already there' (Barthes, 1993, p. 55). The punctum marks the moment that 'I animate the photograph and it animates me' (Barthes, 1993, p. 59). This ambiguity is characteristic of Barthes when he deals with the punctum. He frequently speaks of it as something that can be pointed out in the photograph, something objectively given and there to be seen by everyone, and he actually points out what, according to him, is the punctum in many photographs. However, in the end he leaves out precisely the image that for him is the quintessential photograph, namely the photograph of his mother as child in the Winter Garden, saying:

> I cannot reproduce the Winter Garden Photograph. It exists only for me. For you, it would be nothing but an indifferent picture, one of

> the thousands manifestations of the 'ordinary'; it cannot in any way constitute the visible object of a science; it cannot establish an objectivity, in the positive sense of the term; at most it would interest your studium; period clothes, photogeny; but in it, for you, no wound. (Barthes, 1993, p. 73)

This suggests that punctum is not an objective element in the image after all, but something that exists only in relation to a particular seer, a seer to whom it appears as such. Hence, the Winter Garden Photograph reveals as much about Roland Barthes as it does about his mother. And seen in this way, not showing it, means that he avoids being exposed as a subject. It protects his position as an absent seer who can see it 'as it is'.

6
Retheatricalizing Sexuality in the Field of Vision

'The trap of the visual field' Peggy Phelan observes, 'is that it seems to promise to show all, even while it fails to show the subject who looks, and thus fails to show what the looker most wants to see'. In fact, '[t]he looker is the "not-al" which is left out of the promise of visual plenitude'. Therefore, '[s]eeing is a (false) assertion that the world can be mastered by the gaze *and* a recognition of the world without one self' (Phelan, 1993, pp. 24–25; italics in the original). With this observation, Phelan argues against too optimist expectations about visibility politics. The binary opposition of visibility as power and invisibility as impotence or absence of power, is falsifying. There are serious limitations to visual representation as a political goal. As Phelan points out: '[if] representational visibility equals power, then almost-naked young white women should be running Western culture' (Phelan, 1993, p. 10). The ubiquity of their image, however, has hardly brought them political power.

Visibility does not necessarily serve the interests of the one made visible. Representation is almost always on the side of the one who looks and not on the side of who is seen. The images of women in films, advertisements, in paintings or on stage, do not so much show women 'as they are', as they are revealing about the subjective point of view from where these women are seen. This point of view is outside the image, yet implied within it, like the vantage point in a perspectival drawing. It is the position of the eye of the male artist in Albrecht Dürer's woodcut *Der Zeigner des liegenden Weibes* (1525, Figure 6), firmly locked in its place by the erect pillar that fixates his look on the woman in front of him. It is not difficult to see why this woodcut presents such a rewarding object for twentieth-century feminist critique of the gender bias at work in so-called objective vision.

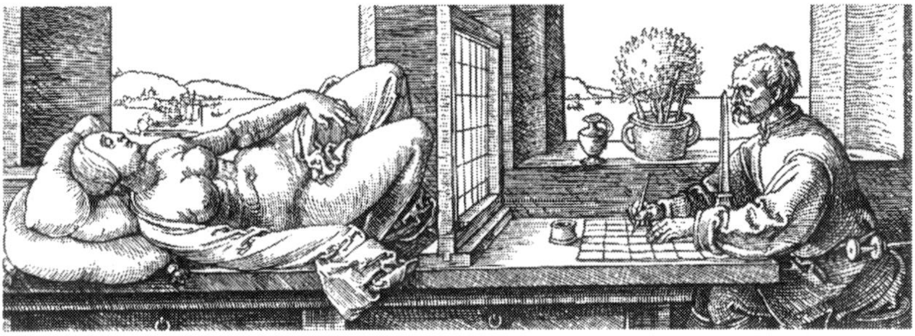

Figure 6 Woodcut *Der Zeigner des Liegende Weibes* (The Draughtsman of the Reclining Woman) (1525) by Albrecht Dürer. Reproduced with permission of bpk / Kupferstichkabinett, Staatliche Museen zu Berlin. Foto: Volker-H. Schneider.

The point of view as implied within the construction of the perspectival drawing mediates in the relation between the woman depicted and the one seeing her. In the Dürer image this mediation is clearly visible and this makes the image such a neat illustration of the paradox given within perspective, the paradox that *is* perspective. On the right, one sees the construction used to produce the image. This construction fixes the eye on one very specific point. By taking up this position implied with the construction of the image, the image in its turn suggests a vision of the world 'as it is' independent of any particular observer.

Perspective presents a model of how representation is firmly on the side of the one who looks, and not on the side of who is seen. This model, furthermore, illuminates how it is invisibility rather than visibility that equals power in the field of vision. Perspective 'theatricalizes' the field of vision. It creates a 'scenographic space' in which all that is seen is in a sense staged for a viewer. At the same time, this staging aims at an effect that is quite the opposite of being theatrical: the promise presented by perspective is one of directness, immediacy, it is the promise of Alberti's *finestra aperta*.

Today, rather than promising a *finestra aperta* on the world, the linear perspective of the early Renaissance is clearly visible as a technique of producing images. Moreover, this is a technique that can be localized in time and place, that can be historicized. Having become all too visible, the technique of linear perspective has lost its power to produce convincing representations of 'how it is'. At the same time, however, this visibility has also increased the power of perspective, not as a means to construct visual images, but as a means to deconstruct them. The explicit relationship between image and viewer as mediated by the point of view given in linear perspective can be used as a model for a retheatricalization of the field of vision. Used this way, perspective can contribute to an understanding of that what remains unseen within the visible: the unspoken, or what Phelan calls the unmarked. The unmarked appears as a blind spot within the visible real, a blind spot that has to remain blind in order to produce the illusion of visual plenitude; that 'inconspicuous allusion which involves both to stage and to hide the subject' (Bal, 1996, p. 176).

Semiotic disruptions

Within the 'inconspicuous allusion' of visual plenitude, blind spots are like the vanishing point in a perspectival drawing. The vanishing point mirrors the vantage point, the point from where the scene depicted is seen. In the Dürer woodcut the vanishing point mirrors the eye of the male artist. The vanishing point marks a location within the scene depicted and at the same time it does not, for it signifies a location that by being infinitely far in the distance is unoccupiable by a person or object. Its meaning can not be understood from what is depicted, but only be retrieved from the process of depiction itself and from the address presented to a viewer and how this address serves the one who looks. Brian Rotman (1987) points out the analogy between the status of the vanishing point in perspective, that of origin as it functions in metaphysics, and of zero in arithmetic. What is exceptional about all three of them is their dual semiotic character. Internally, they are sign among signs, number among numbers. Externally, however, they mark moments of semiotic disruption. All three are meta-signs whose meaning is to indicate, via the syntax that accompanies them, the absence of other signs. In other words, they stand in for the absence of signs, but in doing so they remain signs and, as signs they owe their position to a particular syntax that grants them this status (Rotman, 1987, p. 19).

Arithmetic, perspective and metaphysics, all partake in the constitution of a world as though objectively given outside our systems of representation, while at the same time, these systems themselves are constitutive of this world 'as such'. Within these systems, zero and the vanishing point, like the notion of origin in metaphysics, have a special status: they keep the differences within the system in place by means of a reference to something (supposedly) prior to the system, yet it is the system itself that grants this term its status as prior. As a result of this special status, the vanishing point serves as meaningful pointer drawing attention to the invisible logic at work in the system of which it itself is part.

Facing lack

The vanishing point that is so modestly veiled in the Dürer woodcut and presented through the intermediary of a fictional artist, reappears in a much more explicit way on Courbet's painting *L'origine du monde* (1866, Figure 7) showing a woman's torso, the inside of her thighs, her genitals and a single breast, all in extreme foreshortening. The title of this painting links it to the metaphysical notion of origin; not just some origin but the origin of the world, the origin of human life, the vanishing point par exellence. And what a cunning translation of the invitation implied in the perspectival drawing to 'step inside'.

L'origine du monde represents, as the catalogue of the 1988 exposition *Courbet Reconsidered* puts it, 'the classical site of castration anxiety as well as the ultimate object of male desire' (Faunce and Nochlin, 1988, p. 176). Faunce and Nochlin explain the appeal presented by the painting to the private individual viewing it, as well as the anxiety or shock it frequently provokes, in terms of male subjectivity as it is accounted for in Freudian psychoanalysis. They call the painting 'pornographic' and,

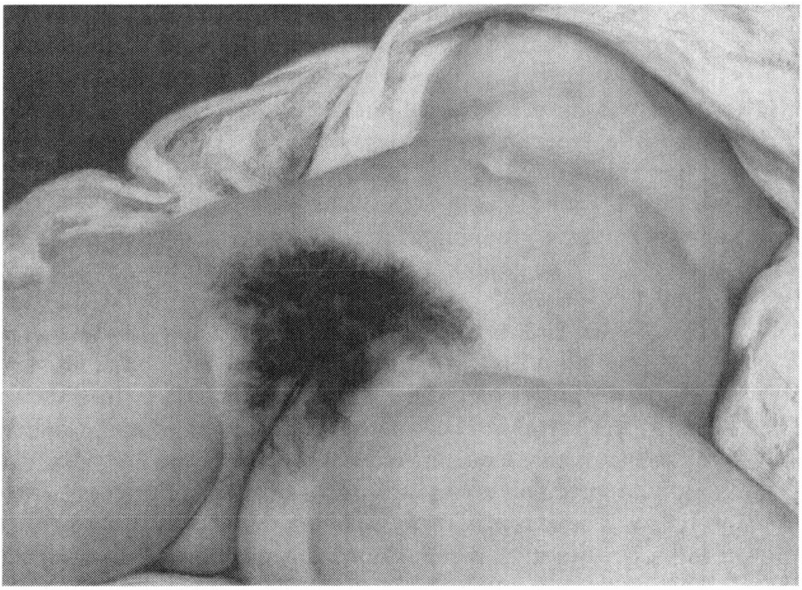

Figure 7 *L'origine du Monde* (oil on canvas, 1866) by Gustave Courbet. Location: Musée d'Orsay, Paris.© RMN. Photo by H. Lewandowski.

indeed, the overt visibility of the site of both castration anxiety and desire plays into what Gertrude Koch describes as the core of pornography: 'the persistent voyeuristic mania to look at the female organ, constantly and as closely as possible, in order to uncover the secret of the missing penis' (Koch, 1989, p. 24). The adult viewer of pornographic images seeks a confirmation of this childhood sexual theory – the phallic myth of the female organ. The phallic myth is Freud's story of the male child who sees the female sex organ for the first time and is amazed that no penis is attached. Disturbed by the fact that an object so important to him is missing from the female body, he imagines a number of equally anxiety-laden possibilities: either the female organ is the result of castration, or the woman is hiding her penis. The second possibility is already a working through of the fears aroused by the possibility of the first possibility. Hence, castration anxiety is both the effect of what is seen, namely absence, and the drive to compulsively view the absence.

L'origine du monde was intended for the private delectation of a sophisticated connoisseur. For a long time, this painting was only viewed by a few and in private. Today, on public display in a world famous temple of High Art in Paris, it draws huge crowds of visitors. It seems that everybody wants to see this painting, but at the same time it appears to be difficult to look at it, much harder than looking at the other representations of naked female flesh as they cover the walls of the Musée D'Orsay, and flank this image. The ambiguous response of visitors of the Musée D'Orsay is subject of Renée Kool's *Looking at L'origine du monde* (provisional title, work in progress) based on a series of photographs taken in the gallery.

At the time I visited Musée D'Orsay to see *L'origine du monde,* the painting was part of a grouping of four paintings that filled an entire wall of the gallery. The grouping was similar in many ways to the one photographed by Renée Kool (Figure 8), with the exception that the painting on the left, showing a naked woman with a dog (*Femme nue au chien,* 1862), then hung on the right side. The place on the left side was occupied by *Le sommeil* (1866), showing two women asleep, their naked bodies draped on a luxuriously decorated bed in a complex embrace that allows the spectator an unencumbered view of breasts and buttocks. The large painting hanging on top of these three smaller paintings is *La nuit de printemps, combat des cerfs* (1877), showing what looks like a herd of rutting deer. Within this composition, the position of *L'origine du monde,* mirroring the position of the eye of the visitor, marks the position of the vanishing point in linear perspective. The female sex organ appears as the lack or sign of absence analogous to the vanishing point

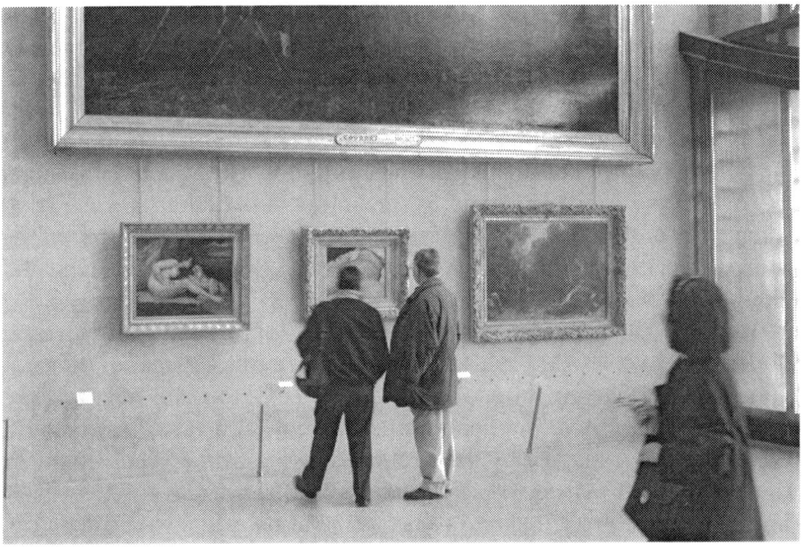

Figure 8 *Looking at L'origine du monde* (provisional title, work in progress) by Renée Kool. Reproduced with permission of the artist.

in perspective, which structures a particular vision, consisting of a constellation in which female flesh, animals and nature are conflated and presented as objects to the vision of the subject whose position mirrors that of the vanishing point represented by the female sex.

Shortly after World War II, Jacques Lacan was the lucky owner of Courbet's painting. Faunce and Nochlin point out how in a history of French psychoanalysis dealing with Lacan's life and work, *L'origine du monde* is described as half of a 'strange diptych', a 'double' painting, consisting of Courbet's painting covered with a wooden 'hiding device', constructed by artist Andre Masson, which represented elements in the first painting in abstract form. A secret system allowed the viewer to slide off the wooden protective covering to reveal the painting beneath, thus allowing Jacques Lacan, to stage the relation between the seer and the female sex organ seen at wish (Faunce and Nochlin, 1988, pp. 176–177).

Faunce and Nochlin also point out the strange analogy between this 'diptych' and Marcel Duchamp's final work *Etant donnés* (*c.*1946–1966) in the Philadelphia Museum of Art. This work consists of a life-size diorama behind a large wooden door with holes bored in it through which the viewer is invited to peep. Through these holes, the viewer sees a brick wall with a hole in it. Behind the hole lies a female body with spread legs and one outstretched hand holding a lamp. She lies in a diagonal angle in a field. Her outspread legs and exposed vulva are the nearest points in this figure to the opening in the wall. The field in which she lies extends in a landscape so foreshortened that it does not have a vanishing point. 'Through this foreshortening the spectator is made aware that the vanishing point in this image is to be found somewhere other than on the horizon' observes Kaja Silverman dryly in her discussion of Duchamp's diorama as part of her critique of Lacan in *The Threshold of the Visible World* (1996).

Duchamp's installation stages the act of looking at a three dimensional version of *L'origine du monde* and explicitly relates the 'pull' presented by the vanishing point in perspective to male sexual desire. Seeing is represented as a penetrating movement directed towards the vanishing point marked by the sex organ of the woman on display. All the elements of perspective are in place, but in a strangely literal way. The suggestion presented by the perspectival painting, that the picture plane opens up to a space beyond, is staged here as a matter of literally breaking through the barrier presented by the brick wall. The ragged edges of the hole in the wall seem to indicate violence. Duchamp's staging of the vanishing point behind a hole in the wall is a reminder of many Renaissance painters' habit of placing the vanishing point inside a hole: a framed opening such as a door, a window, a mirror or even another painting, as if to emphasize the essential otherness and exteriority of its location. This had the effect of doubling the pull exerted by the vanishing point on the spectator and, at the same time, of pushing the vanishing point out to an infinite, unreal, numinous distance by invoking the potentially unlimited iteration of a frame within a frame (Rotman, 1987, pp. 18–19). The scopic field in which woman and landscape conflate, is literally 'feminized', passive, accessible, turning the woman, spread out across the landscape, into an emblem of the scopic field itself. The explicit staging of Duchamp's piece, highlights the relationship between seer and seen as given in perspective. The retheatricalization exposes the address presented to the viewer and undermines the absorptive qualities of the invitation to 'step inside'.

Duchamp's installation features prominently in both Rosalind Krauss's (1993) and Kaja Silverman's (1996) critiques of modernist notions of disembodied vision. Both remind us of Lyotard's discussion of Duchamp's work in *Les TRANSformateurs Duchamp* (Paris: Galilee, 1977). Krauss points out how, for Lyotard, the meaning of *Etant donnés* is to be found in the way it comments on the implicit equation of the vanishing point and the point at which the observing subject is encouraged to stand before a perspectival image:

> In this type of organization, the viewpoint and the vanishing point are symmetrical. Thus if it is true that the latter is the vulva, this is the specular image of the peeping eyes; such that when these think they are seeing the vulva, they see themselves. *Con celui qui voit.* He who sees is a cunt. (Lyotard quoted in Krauss, 1993, p. 113; italics in the original)

For Lyotard, the significance of Duchamp's installation is that it works to corporealize the male eye. In this incarnation, the female body appears as the object of a vision that originates not from the body seen 'over there' but rather whose origin is located in the male eye. Silverman, in her comments upon both Duchamp and Lyotard, points out that the installation also effects a remarkable deconstruction of the grounding opposition upon which sexual difference is based:

> The male look at the female genitals is emblematic of that exteriorizing displacement through which the male subject repeatedly situates his lack at the site of the female body, and naturalizes it as essentially (i.e. anatomically) 'other'. The additional oppositions which the Duchamp diorama works both to articulate and collapse – carnality/transcendence, spectacle/eye – are built upon this foundational antithesis. When the male voyeur who is caught within the machinery diagrammed by Lyotard [...] finds his look put in a mirror relation to the female genitals, he is given the opportunity to acknowledge as his own what he is accustomed to throw violently away, and so, to renegotiate the relation between his ego and the object. (Silverman, 1996, p. 172)

Mirror, mirror on the wall, the seer wants to see it all, but what he sees is absence or lack. This mirror, instead of reflecting back an ideal image, invites the viewer to identify with lack. Instead of reflecting the visual plenitude promised by the perspectival construction that the installation mimics, it undermines the position of the seer as the one who can see it as it is. Duchamp's installation positions the viewer literally at a peephole, thus exposing the voyeurism involved in perspectival vision. The place of his installation in the museum, a public place, undermines the position of the seer as neutral, absent, non-desiring, disembodied eye.

> Threatened by discovery on the part of a fellow viewer, the purely cognitive subject of Kant's aesthetic experience is redefined in this

> setting as the subject of desire, and subjectivity itself is taken from the faculty of cognition and reinscribed in the carnal body. (Krauss, 1993, p. 114)

Classical perspective orients the field of vision to the viewer's invisible body, as if the scene seen had emanated from the viewer's own eyes, while at the same time, this eye is erased from implication within the visual field. The degree to which the seer is invisible, detached, is the degree to which he bears authority within the terms of perspective, but it is also precisely the degree to which he is vulnerable to being caught seeing. To be caught seeing is to lose one's prerogative as a disinterested viewer. To be caught seeing is to be robbed of one's privilege and power as seer; it is to be castrated, to be turned into an object of vision. To be caught seeing is to be feminized.

Con celui qui voit. He who sees is a cunt. In Renée Kool's photograph, a male viewer literally becomes the cunt he is watching when his hair fuses with the pubic hair depicted in Courbet's painting. His legs, spread like the legs in the picture he sees, turn his body into a physical mirror of the painting, an act of physical mirroring through which he becomes the female figure. What this photograph also demonstrates is that all of this happens within the 'vision' of a third party. The fusion requires a perspective within which all of this takes place. This third party is represented in the picture by the woman on the right side, watching the men, but even more by the photographic apparatus that caught this man seeing.[1]

Echoes of Duchamp's staging of the seer as carnal body both seeing and seen re-appear in Annie Sprinkle's performance *Post Porn Modernist,* in which she invited the audience to look inside her genitals through a speculum (Figure 9). In this performance, the artist places her own body on display. This is not some objectively given body (the female body) but this individual body here and now. It is a body furthermore, that looks back. Sprinkle literally puts the audience and their act of looking on stage and thus exposes these seers to the look of others (including Sprinkle herself) looking at them. In this way, her performance undermines what perspective grants the beholder: a safe position marked by absence–an absence that is the very condition of the privilege of the seer as a disembodied observer.[2]

Yet more echoes of Duchamp's staging reappear again in Gonnie Heggen's dance/theatre performance *Looking for peter* (1996),[3] in which she draws attention to the a-symmetry at work in the way signs of sexual difference are perceived as a means of exposing the subjective perspective involved in 'just looking' as it takes place in the theatre. She does so

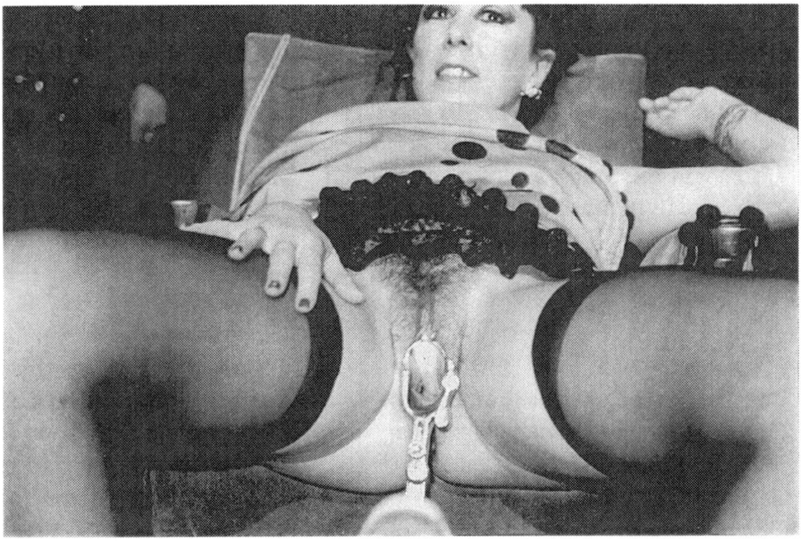

Figure 9 Annie Sprinkle's *Public Cervix Announcement* from her *Post Porn Modernist* show performed at the Kitchen, New York (1990). Reproduced with permission of the artist.

by means of a theatrical commentary on James Barrie's famous children's play *Peter Pan*. Barrie's play tells the story of three children – Wendy, Michael and John Darling – who, one evening, when their parents have gone out, receive an unexpected guest named Peter Pan. Together with his fairy Tinker Bell, Peter Pan is looking for his shadow, which he finds in a drawer in the Darling nursery. Wendy helps Peter to sew the unruly shadow on again and in return Peter teaches the Darling children to fly. He takes them with him to Neverland, a fairy tale place beyond the stars where he lives amidst pirates, Indians, mermaids, and other fairytale figures. They have exciting adventures until Wendy decides it is time to go home because their mother might get worried. Before they can actually leave, they are kidnapped by the pirates, but Peter Pan manages to defeat Hook, the captain of the pirates, and they returns home safely. He himself returns to the fairy tale world of Neverland, where time has no hold on him and he can stay a little boy eternally. First performed in 1904, Peter Pan was an instant success, especially in Great Britain and the United States, where the play would keep the stage for decades and become inextricably bound up with the Christmas season. Several film versions, including Disney's famous cartoon, and a popular musical version of the play, added to its fame.

Heggen's *Looking for peter* is not a staging of Barrie's play, but a commentary on the way in which this play mediates in looking at sexual difference. She does so starting from a remarkable aspect of the tradition of staging this play, and this is the fact that Peter Pan is traditionally played by a woman. With her performance, Heggen too, holds up a mirror to the spectator, confronting him or her with his or her own looking. Unlike Duchamp's re-staging of *L'origine du monde*, this mirror does not present the spectator with a clear vision of otherness, but rather confronts the spectator with its invisibility.

Back to nature

'Peter Pan is the ultimate ambition of all actresses just as Hamlet is of all actors', writes Roger Lancelyn Green in 1954 in his book *Fifty Years of Peter Pan* (quoted in Garber, 1992, p. 165). There is, as Majorie Garber points out, a curious absence of symmetry in this apparently symmetrical assertion, since Peter Pan, as well as Hamlet, is nominally gendered male. Yet unlike Hamlet, Peter Pan has traditionally always been played by a woman. Hamlet too has been played by women many times, some of them the greatest actresses of their time. Nevertheless, it seems highly unlikely that Hamlet would ever be called the ultimate ambition of every actress. But Peter Pan is, if we are to believe Roger Lancelyn Green.

In 'Fear of Flying, or, Why is Peter Pan a Woman?', Garber (1992) reminds one of the rare occasions a major professional company cast a man in the role of Peter Pan. This was with the Royal Shakespeare Company in 1982. The director explained his move by declaring his desire to make the play a real tragedy. He spoke of 'elevating it from the ghetto of children's theatre into a national masterpiece' (quoted in Garber, 1992, p. 165). And to do so, what he needed was a man. Actually, it is not difficult to imagine that it makes a difference whether Peter Pan is played by a man or by a woman. It is much more difficult, however, to account for this difference and to explain it in terms of signs and referents. When a famous actress plays Hamlet, her stage presence represents the absent fictional prince of Denmark. Her female body functions as a sign for the male body of the prince. The difference between sign (the female body of the actress) and referent (the male character Hamlet) can contribute to the meaning of the performance, or to the pleasure derived from seeing it. If, though, one wanted the audience to believe that this person he or she is seeing here and now is Hamlet rather than representative of Hamlet, that is, if one were to strive to achieve the effect of apparent conflation of sign and referent on stage, casting an actress as Hamlet does not seem a very productive choice. Nevertheless, Peter Pan is played by a woman exactly because this supports the conflation of sign and referent, even though the male gender of Peter Pan would seem to suggest the opposite.

In Gonnie Heggen's performance, Peter Pan is ambiguous in appearance possessing both male and female features. Dressed in a camouflage shirt and shorts, with blond dreadlocks and a baseball cap, he/she frolics happily through a landscape inhabited only by animals. This Peter Pan has a mission: the god given task to measure with a male/female indicator the male and female energy in all living creatures in order to give 'him up there' an impression of how things are going 'down here'. The character refuses to be classified as either male or female. He/she argues that he/she does not fit into the male–female binary because, he/she says that in him/her, male and female energy are completely in balance. He/she is neither male nor female and both at the same time. With his/her ambiguous sexuality, Peter Pan seems to represent a third possibility: a third sex that is neither male nor female. 'In nature everything is both male and female', says Peter Pan, and presents his/her 'natural' condition as a more authentic state of being, a presumed state of original fullness that we can see reflected in many stories of paradise or Arcadia: 'the utopian ideal of the world as a garden in which human bodies are so at one with their environment that they lack nothing' (Halkes, 2001, p. 14; see also Chapter 4). These stories are lodged within the imagination of Western culture as a utopian ideal, as a 'recovery narrative,' as a sustained attempt to recover a lost ideal of the body, wanting in nothing in the garden of plenty.[4] This presumed state of original fullness is also reflected in the psychoanalytical concept of the subject.

For the subject of psychoanalysis, the desire for fullness and completeness manifests itself in the desire for the penis (Freud) or the phallus (Lacan). In Freud's account, men appear as those who have, who are complete, and women as those who are incomplete and want to have what they lack. Penis envy appears as the very cornerstone of female subjectivity. It marks the castration complex of the young girl: her wish to be able to exhibit the penis she does not have. Successful maturation will convert this female wish for the penis into the wish for a baby. Ultimately the outcome of the infantile wish for a penis becomes the wish for a man, prompting the woman to accept a man as an appendage to the penis.[5]

Lacan, in his reading of Freud, replaces the centrality of the penis with the centrality of the phallus. The phallus is not the penis, it is not an organ, it is a signifier. Lacan uses the term phallus to designate all those values that are opposed to lack. In Lacan's account, lack defines both male and female subjectivity. As in Freud, this lack is sexual in definition: it starts from the impossibility of being both male and female. One of the assumptions that underlies the Lacanian notion of the subject, is that the subject derives from a primordial whole, that it becomes a subject by separating from this whole, and that this separation inaugurates lack as well as a desire to return to primordial wholeness.[6]

The character Peter Pan defines him/herself as a being without the lack that defines the subject of psychoanalysis because he/she successfully refuses to submit to the division that inaugurates lack in the first place. Although at first this may seem to represent a successful return to nature, in the end Peter Pan's authentic condition is shown to be a cultural projection. In *Looking for peter*, the character Peter Pan literally only exists as projection, as part of a film that is projected onto a screen framed by antlers and pieces of bark. In this film, Peter Pan appears, as a critic described it: 'a cheerful, esoteric creature of the woods that has taken a bite of all religions without swallowing anything so it can, in a carefree way, shed confusion on all that is blossoming and flowering'.[7]

Peter Pan's encompassing condition is shown to be a projection screen for a desire for wholeness that, as Garber points out, at the time of her writing found its most prominent expression in popular culture in the figure of Michael Jackson. With his ambiguous identity–neither man nor woman, child nor adult, black nor white, he seems to join oppositions in a union without tears. He suggests the possibility of a conflictless unification of differences by being one and the other in perfect balance. His public imago includes the promise that with him, the m/f indicator will hold exactly in the middle. Like Peter Pan, Jackson's appearance is sex-less rather than sexually ambiguous (in contrast to the ambiguity involved in transvestism which is usually erotically charged). Indeed, his childish innocent appeal is so strong that his image has even managed to survive serious accusations of child abuse.

With their successful refusal to grow up, both Michael Jackson and Peter Pan have placed themselves outside society. They can only exist in a fantasy world. In *Looking for peter*, this fantasy world is 'nature': nature that is shown to be a product of culture, a kitsch illusion populated by animals that upon closer look appear to be either already dead and stuffed, or domesticated and living in the zoo. Michael Jackson too is a cultural construct, even at the most physical level. The oppositions he unites in his appearance are not kept together from the inside. Instead, unity has to be won over and over again by technical means against the very serious threat of falling apart. Like Peter Pan, Jackson has fled the world of the grown ups. In a far away place, he has created his own fantasy world where he spends his time in a private playground named *Never Land*.

To have or not to have

Peter, apart from being the name of the famous character, is also slang for the male member. In Freudian psychoanalysis, 'looking for peter' marks the crucial moment in the development of sexual identity. In 'Some Psychical Consequences of the Anatomical Distinction between the Sexes', Freud describes this critical moment in infantile sexuality that establishes the concepts of maleness and femaleness and determines a very different history for each. It is a moment in which vision plays a central role, in which the male and female subjects first 'see' each other (Freud, 1961a).

According to Freud, it is not becoming aware of the anatomical differences between penis and vagina that does the job. The awareness of sexual difference results from 'looking for' the absence or presence of one single factor. The difference between man and woman thus appears as the difference between having and not having, between man and not man. That is, it is not the difference between two sexes, but between one sex and its absence. In the Freudian account, the female genitals stand for the absence that inaugurates the awareness of sexual difference, and thus inaugurates a fundamental binary opposition.

In this view, women appear as neutered in contrast to men, who are marked by a sign that defines their sexuality. It seems logical then that a woman would be better suited to play Peter Pan than a man. Who could be more suited to play a boy who does not want to grow up and become a man than a woman, a non-man, someone who will never be a man? As Garber observes, given this, the choice of the Royal Shakespeare Company to cast a man in the role literally means putting the 'peter' in Peter Pan, a critical move that, in the eyes of this company, would take the play from the backwaters of kid's stuff and turn it into a national masterpiece of tragic drama. In *Looking for peter*, Gonnie Heggen presents her own version of what it means to put the 'peter' in Peter Pan, and this does not turn her performance into a tragedy, quite the contrary.

Gender as performance

On stage, four women dressed in brown jackets and tight fitting trousers demonstrate how they are able to transform themselves into 'real men,' as 'real' as we rarely get to see them. They cross the stage with sturdy steps, spit high into the curtains and burp into the microphone. Just as easily they transform into 'real' women. Under their jackets they wear colorful, seductive tops. They trip around on staggeringly high heels while looking at the audience with a helpless expression on their faces. Seemingly effortless they transform into various animals as well.

With their performance, these women do not represent a story but show transformations in looks and behavior, constantly crossing the lines dividing the sexes and dividing culture and nature. The movements suggest recognizable patterns but before they are fully developed they have already transformed into something else. The behavior of the women on stage blurs the boundaries between male and female, nature and culture, and human and animal, as categories ontologically grounded in what is 'over there'. They show all of it to be 'just performance' depending on using the right signs. Seen in this way, they present an illustration of Judith Butler's dictum underlying her *Gender Trouble* (1990), that gender is performance. They also illustrate that this performative aspect of gender does not imply the freedom to do and be at will.

The theoretical concept of gender as distinct from sexuality has proved useful in understanding the ways sexual identity manifests itself as culturally mediated rather than natural and given. From the perspective of gender, sexual identity is not the destiny implied within one's anatomy but a cultural construct. The distinction between sex and gender has helped to criticize roles and ways of behaving as granted to men and women at a particular time and place. It is instructive to understand sexual identity as a role whose traits are not naturally given qualities inherent in bodies but culturally determined functions that manifest themselves in behavior which defines the actors involved and not vice versa. The distinction between sex and gender has contributed to a more complicated conception of sexual identity. At the same time, however, this distinction runs the risk of turning into a binary opposition once again, an opposition that repeats the old nature/culture binary in which sex is understood in terms of materiality as biologically given, and therefore, more natural than gender as 'just' cultural performance, inscription, overlay. It is this opposition that Judith Butler argues against in *Bodies that Matter* (1993), which followed *Gender Trouble*.

In *Bodies that Matter*, Butler argues that although matter is posited as the neutral, non signifying outside of signifying systems, this happens within a logic in which matter is associated with the feminine and signification with the masculine. Within this logic, materiality appears as the effect of signifying practices that produce it as 'other', as that which falls outside signification. Matter and materiality, Butler argues, appear as neutral, that is, as opposed to signification, only from within this gendered perspective within which the absence of signification and the feminine are conflated. This perspective is what is at stake in *Looking for peter*.

Productive looking: the penis as prop

At first, Heggen's performance only seems to confirm the idea that gender is like a set of clothes one can don at will. The audience knows that the bodies of the dancers seen on stage are female, yet we also see how they convincingly transform into men and animals by giving signs that make us understand them as such. This would suggest that the body, and more precisely, the female body, is a neutral ground for inscription through a performance that turns its materiality into a meaningful appearance.

But then, at one point, a macho demonstration of masculinity smoothly transforms into a striptease. Each of the women brings out a powder puff, traditionally a female attribute that heightens the femininity of the one using it or seen with it, in the same way that breasts (fake and otherwise), long lashes, long hair, high heels and corsets heighten the femininity of the one seen with them, even if this one is a man, as in the case of transvestism. These attributes are understood to function on a cosmetic level. They do not really change the sexual identity of the one wearing them. They are part of the signs used in gender performance. When a man wears lashes or false breasts it does not turn him into a woman. Yet it does make him look more female. Similarly, many kinds of surgical interventions are understood to function on this cosmetic level. A breast job does not turn a man into a woman. There is only one operation that changes sexual identity.

When in *Looking for peter*, the women on stage put their powder puffs into their panties to use them as signs of masculinity analogous to the way fake breasts can serve as signs of femininity, this is indeed a hilarious moment (Figure 10). They effectively present the 'powder-puff-as-sign-for-masculinity' as an exchangeable signifier, a stage prop, but what their performance demonstrates is that it does not work that way at all. While for a man, masking the sex organ may contribute to his appearing more feminine, for a woman, hiding the female sex organ behind a fake penis does not make her look more masculine. The penis as a sign for masculinity appears to function unlike other signs. In order

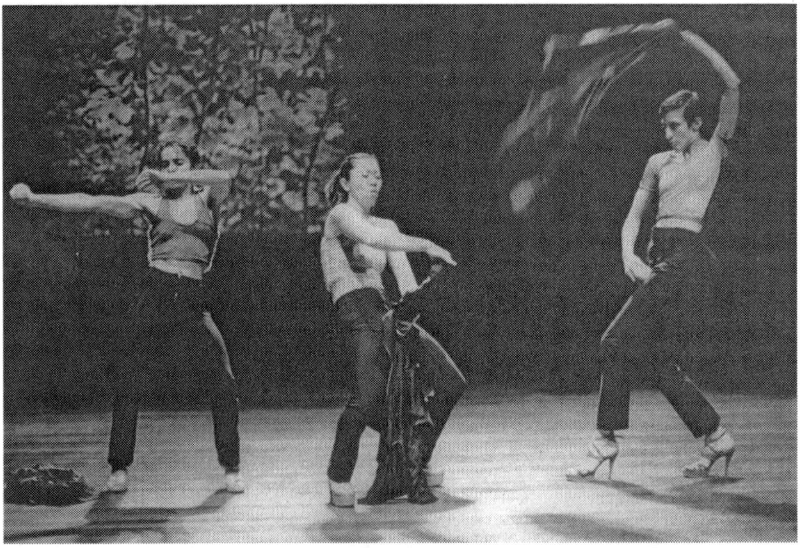

Figure 10 Photograph by Maurice Boyer showing Vivianna Rodriguez de Brito, Yasuko Yokoshi and Gonnie Heggen in Gonnie Heggen's *Looking for peter*. Reproduced with permission of the photographer.

to function as a sign of masculinity, the penis must be attached to the male body; it must be part of it. If it is not, it signifies the absence of masculinity instead. The penis, as a sign of masculinity, thus appears to depend upon a situation in which sign and referent (seem to) be conflated. That is, it depends upon a situation in which the penis does not appear as a sign at all, but as the full self-validating presence that evokes the 'there it is' quality described in the previous chapter.

By presenting the 'peter' literally as a sign, Gonnie Heggen performs a deconstructive reversal of the binary opposition that inaugurates dominant conceptions of sexual difference. She shows how this binary opposition results in a subjective deviation in the visual field. She demonstrates that an alternative perspective is thinkable. From the point of view of the female body, this body is not lacking anything. The male body then is a body with something extra, a part that functions as a sign. If it is present, the body signifies male. If it is absent, the body signifies as not male. Heggen's performance also demonstrates that this subjective deviation cannot simply be undone and that we cannot simply choose to see it one way or the other. The different perspectives presented are not equally visible. Yet, her performance does make a change in the sense that it denaturalizes the dominant perspective and invites what Silverman (1996) has termed *productive looking*.

Silverman acknowledges that it may very well be that there is nothing we can consciously do to prevent certain projections from occurring over and over again in an almost mechanical manner. That does not mean though, that it is not possible to make a change or, as she calls it, 'to make the ethical operative':

> The ethical becomes operative not at the moment when unconscious desires and phobias assume possession of our look, but in a subsequent moment, when we take stock of what we have just 'seen' and attempt – with an inevitably limited self-knowledge – to look again differently. Once again, then, the moment of conscious agency is written under the sign of *Nachträglichkeit*, or deferred action. (Silverman, 1996, p. 173; italics in the original)

It is this reconsideration, this becoming aware of the relationship between seers and seen, that can turn looking into productive looking, and that can turn confusing experiences on the post-dramatic stage into a political act.

According to Silverman, one way to make the ethical operative is theatricalization. Theatricalization can be used to show bodies that, within the current symbolic order, appear as non-ideal in the guise of the ideality they lack. At the same time explicit theatricality helps to understand this ideality 'as a garment rather than the body itself' (Silverman, 1996, p. 103), that is, it helps us to understand the appearance of bodies as the effect of signs, signs we read and interpret, and that get their meaning within a frame of reference, rather than as something essentially given within them. The staging of God and His companions in *Klaagliederen* (as discussed in Chapter 3) might be called such a moment that aims at evoking productive looking. In *Looking for peter*, the theatricalization of the penis, its appearance as sign rather than given, confronts the audience with a process of framing as it is at work within its own looking: where is the border between what you think you see and what is actually there to be seen? Is it possible to see the difference between the powder puff bulge and the bulge of the male member? What is actually there to be seen and where do your own projections, desires, and denials begin? Constantly shifting their behavior between male and female, human and animal, the dancers' performance questions distinctions between nature and culture, sex and gender, what is there to be seen and what is produced by our culturally determined look.

The performance does not provide the audience with a clear frame through which to understand what is there to be seen on stage. Instead, the performance highlights ambiguity as we seem to be able to see everything 'as it is' and yet feel that our vision is somehow 'framed', that we are not really free. *Looking for peter* makes us as seers aware of perspective as something within which we are implicated and that guides our looking, as though someone was looking over our shoulder and directing what we see and don't see. This 'someone' shows us the world according to an invisible logic, and since this logic will determine what becomes visible and how, a special strategy is needed to expose this agency. Gonnie Heggen manages to do so by means of a strategy that Majorie Garber has called *empowered transvestism*.

Empowered transvestism

Garber explains the meaning of her notion of empowered transvestism with the help of a video clip by Madonna. In the clip that accompanied the song *Express Yourself*, Madonna imitates Michael Jackson. Dressed up in a double-breasted suit and Jackson's famous white socks and shiny black men's shoes, she dances in a style directly imitative of Jackson, mimicking many of his moves including his trade mark crotch grabbing. When performed by Jackson, this move has not been considered offensive or objectionable. Performed by Madonna, it gave rise to controversies and was considered to be obscene.

> Madonna is a famous female star who is impersonating a famous male star who is celebrated for his androgynous looks and his dancing style. Why is it shocking when she grabs her crotch, repeating as she does so a gesture so familiar to anyone who has watched a two-year-old male child reassuring himself of his intactness? Not because it is unseemly for a woman to do this – although it may be so, to some people – but because what she is saying, in doing so, is: I am not intact, he is not intact; I am intact, this is what intact is. (Garber, 1992, p. 127)

In the case of Jackson, the challenged gesture can be understood as a confirmation of what he has. Madonna derives her pleasure from squeezing what she hasn't got, or has she? She imitates Michael Jackson's looks and behavior and demonstrates how this masquerade gives her pleasure. This pleasure derives not from pretending to have what she has not got. She does not pretend to be Jackson, she performs like him. Neither is what she is doing a female masquerade aiming at seducing someone who has got what she does not have. Instead, she seems to enjoy performing herself as she is.

Analogously, in *Looking for peter* the installment of the powder puff as penis does not happen on the quiet, suggesting an attempt to pass for real. The installment happens front stage, and at arms length from the first row of spectators. While doing this, the dancers look the audience in the eyes as if trying to figure out their response. The dancers show that they know that they show, that they know that they are being watched, and that they like it. Like Madonna, they present their game openly, thrusting their pelvis and seeming to take pleasure in their performance. The audience responds with nervous giggling.

'Of course I lack (the penis)', is what Madonna seems to say. 'So what? That does not mean that my being is necessarily driven by wanting to have (this penis). More than that, it is only from the point of view of someone who has, that I can appear to be lacking and wanting to have. I can only appear as wanting to have from the point of view of someone who is driven by castration anxiety, that is, from the point of view of someone who interprets my being different as lack.'

With their acts of empowered transvestism, Madonna and Gonnie Heggen show the focus on either the penis or the phallus as being opposed to lack and, therefore, to what we all want to have, itself to be a masquerade that serves to veil something else. Their performance suggests that the focus on lack as the driving force behind a continuous desire to get what one does not have, serves to veil the fact that the vanishing point keeping the symbolic order together is not the fear of lack or castration, but the fear of its discovery. Their performances suggest that it is not castration anxiety that keeps the symbolic order together, but the way in which castration anxiety goes unquestioned. This is the real vanishing point hidden behind the continuous reiteration of scenarios that re-enact this childhood sexual theory – the phallic myth about the female organ.

This is not to deny the power of castration anxiety within the current symbolic order, nor is it to deny the usefulness of the concept of castration anxiety to understand the way experience and meaning comes into being in interaction with cultural artefacts such as for example Duchamp's *Etant donnés* or Courbet's *L'origine du monde*. But it is to argue that performances like Madonna's or Gonnie Heggen's point to the need to understand both the Freudian and the Lacanian model of the subject in relation to the discursive practices which defined their immediate context, and which continue to prevail. That is, they point to the need to understand that, what in their models appears as primordial wholeness preceding absence or lack, is not something given, but is itself a product of the symbolic order. Madonna and Gonnie Heggen's empowered transvestism points to the need to understand the perspective within which castration anxiety can appear as such a central term. Whose interests are being served by its unquestioned position?

Although at first it may seem that castration anxiety poses a threat to those who have, and can therefore be castrated, this is in fact not the case. The presupposed link between having and a (supposed) primordial unity turns those who have into the winners. To put it in Freudian terms, the little boy's fear of lack gets projected onto the little girl, turning her state of being different from him into lack and confirming his superiority. Because she is missing what is important to him, he assumes that she is driven by a desire to have what he has. It is only because this vision is then imposed on the girl as well, that she can begin to understand her being different in terms of lack. What is at issue in penis envy is not some biological inferiority. The attribution of superiority and presence to the penis, and inferiority or lack to the clitoris, occurs from within patriarchal culture. It is, as Silverman points out, only after the subject has arrived at an understanding of the privileged status afforded to men and the de-privileged status afforded to women within the current symbolic order, that sexual difference can be read in the way suggested by Freud. It is only retroactively that anatomy can be confused with destiny. This confusion performs a vital ideological function because it serves to naturalize or biologize what would otherwise be open to question (Silverman, 1983, p. 142).

The Lacanian distinction between penis and phallus suggests the possibility of undoing this naturalization, and thereby overcoming the a-symmetry between the sexes. According to Lacan, neither men nor women have the phallus. They both want the phallus because the phallus stands for what they lack. Yet only the Other has the phallus; the subject, whatever organ he or she may have, is symbolically castrated. This 'Other' is an ideal other to whom we contribute what we lack. We are symbolically castrated because we have fallen from a supposed primary condition of being at one with the world. Instead our connection with the world that surrounds us is always mediated through the symbolic order into which we are born. This symbolic order provides us with our ways of making meaning, but at the same time it implies alienation from what is presumed to be our origin and from what might be called the origin of meaning.

In practice, however, the distinction between penis and phallus appears to be more complicated and has been the subject of vehement discussion. The signifier phallus sounds and looks differently than the signifier penis. It produces different associations. But within the current symbolic order it also always refers to the penis. Or as Silverman puts it: 'Despite Lacan's repeated reassertions that the penis is not the phallus, it is clear that there is a very intimate and important relation between the two' (Silverman, 1983, p. 185).[8] Some might wish to polarize penis and phallus into a neat opposition, but as Jane Gallop (1988) remarks, such attempts to remake language fit to one's own theoretical needs, as if language were merely a tool one could use, bespeaks a very un-Lacanian view of language.[9] The desire to control the meaning of the signifier phallus is itself a manifestation of the symbolic castration that sets the Lacanian model in motion. In this model, the presupposed primordial unity functions like the notion of origin in metaphysics, or the vanishing point in perspective. Like the notion of origin in metaphysics, this notion of primordial unity is posed as a primary presence and as such it helps to keep the system of differences within the symbolic order in place. But, as Rotman points out, this primary presence owes its status to the system it is supposed to precede. It can appear as primary only from within the system. Seen this way, the replacement of the penis with the phallus functions like a 'recovery narrative', that helps to preserve the dream of original fullness by projecting it outside the symbolic order. Such a presumed original state is, as *Looking for peter* demonstrates, itself a product of culture, a projection screen. Projected outside the symbolic order it helps to keep the differences within the system in place.

Again, this is not to say that we can simply change the invisible perspective that underlies how the world becomes visible at will. But critical gestures like those made by Madonna and by Gonnie Heggen can help to 'make the ethical operative', by alerting us to aspects of this perspective and how they mediate in what we think we see. This perspective is kept together by the tendency to reduce heterogeneity to a clear-cut opposition between self and other, and a potential multitude of differences to the opposition of plenitude and lack, presence and absence, as well as the many other binary oppositions that structure dominant ways of seeing and ways of thinking, including what is still the dominant model of the subject.

7
Disorders That Consciousness Can Produce

As Duchamp in his *Etant donnés* (discussed in the previous chapter) shows, the seer is always also an element seen within the visual field. This condition of being seen, of being part of the visible world, is what threatens to undermine the condition of the seer as the one who is able to see it 'as it is'. This awareness undermines the position granted by perspective, a position marked by absence. Awareness of being seen undermines the illusion of mastery of the visual field because it brings with it an awareness of being seen from a position from which one cannot see oneself. In this chapter, I will explore some of the implications of this intertwining of seeing and being seen in dialogue with Heinrich von Kleist's text *Über das Marionettentheater* (*On the Puppet Theatre*, 1810), and a performance based on this text titled *The Path of the Dancer's Soul* (*De Zieleweg van de Danser*).[1] The title of this show is a quote taken from von Kleist's text, which consists of three short stories about bodies seen followed by a conclusion that turns the three stories into an argument about the relationship between self-consciousness and bodily grace.

According to von Kleist, self-consciousness and bodily grace are antithetical. The less self-conscious living beings are, the more beautiful their bodily appearance, thus relating grace and bodily beauty to a state of innocence that precedes the awakening of self-awareness. Typically, von Kleist's text directs all attention to bodies seen, implicitly placing the seer in the position of the detached observer. In their staging of von Kleist's text, Rijnders and De Châtel manage to redirect attention and expose the body involved in seeing. Their re-theatricalization of the relationship between bodies seeing and bodies seen, reveals von Kleist's narrator's objective observations on bodies seen as a highly subjective account. Rather than resulting from the consciousness of the bodies he sees, it seems to be his own consciousness that produces the disorders he observes in them.

Mirror, mirror on the wall

In von Kleist's text, a first-person narrator tells about an encounter some years earlier with Mr C, the chief dancer of the opera-ballet in the town of M. Much to his surprise, the narrator discovers that Mr C is a great admirer of the puppet theatre. Mr C appears to think that puppets make much better dancers than human beings, and he goes so far as to state that 'any dancer who wished to improve his art might learn all sorts of things from them' (von Kleist, 1982, p. 211). The narrator sits down with him to listen to the grounds on which he bases this remarkable assertion.

According to Mr C, the movements of the marionette are more beautiful because they are more natural, and he explains this beautiful naturalness, or natural beauty, in terms of a lack of self-awareness. He acknowledges that the range of movements that these mechanical bodies can produce is, of course, limited. But the lightness, grace and serenity with which they are executed must, according to him, amaze every thinking person. These movements are produced by an operation that may seem simple from a mechanical point of view – consisting of a movement of the centre of gravity in a straight or curved line – but which is really quite mysterious, 'for it is nothing less than *the path of the dancer's soul*' (von Kleist, 1982, p. 212; italics in the original). In the mechanical puppet, this soul is located in the exact centre of gravity of the puppet's body, and this is what makes their movements so graceful.

> Unlike living dancers, the puppets have the advantage of counter-gravity. They know nothing of the inertia of matter, which of all properties is the most obstructive to the dance: for the force that lifts them into the air is greater than that which pulls them to the ground. (von Kleist, 1982, p. 214)

This weight that grounds living dancers also appears to have a metaphorical meaning. Mr C associates the disturbed relationship between movement and soul, which he observes in living dancers, with the Fall. Living dancers paid for their bite from the apple of knowledge with the loss of grace. More than their material weight, it seems to be their conscious awareness of themselves as bodies seen, that disturbs the relationship between their soul and their movements. As a result, their vis motrix – or soul – is usually located at any point other than the centre of gravity, which results in the disturbance of natural grace and produces affectation, which Mr C associates with baroque mannerism.

For Mr C, the dancing body of the marionette represents a state of natural grace and innocence – a state that he, as a conscious being, has lost. The naturalness and truthfulness he perceives in the marionette's movement, therefore, do not result from a convincing representation of bodies as he knows them. It is not because he is able to recognize himself in these bodies that they are so attractive to him; on the contrary, the marionette's true and natural movements are more ideal than his own, and they therefore seem to promise fulfillment of what he lacks. Looking at the marionette's dance provides him with a link to a state of innocence he has lost. It provides him with a provisional, temporary, and of course – in the end – incomplete recourse to what he will never (again) be.

The narrator replies that he, too, knows only too well the 'disorders that consciousness can produce' (von Kleist, 1982, p. 214), and tells of an event that happened to him three years before. He had been swimming with a young man whose physical form seemed to radiate a marvelous grace, of which the young man was not yet aware. While the young man was drying himself, he unconsciously assumed the position of a famous statue called the Spinario, a statue they had admired in Paris shortly before. Looking at himself in a large mirror, the young man recognized the similarity and jubilantly pointed it out to the narrator. The narrator, however, although he had had to admit that he had noticed it too, laughed, saying that the boy was seeing phantoms. The boy, confused, tried to re-create the jubilant moment, but was not able to do so.

> From this day on, as though from that very moment, an inconceivable transformation began in that young man. He would stand whole days before the mirror; one charm after the other fell from him. An invisible and incomprehensible force, like an iron net, seemed to spread over the free play of his gestures, and when one year had passed not a trace could be detected of that sweetness which had once so delighted the sight of all who surrounded him. (von Kleist, 1982, p. 215)

In front of the mirror, the body seeing and the body seen, separated in the case of the marionette, merge into one, and it is exactly this merging that, according to Mr C, produces the reprehensible condition of self-consciousness. The marionette's inability to see prevents it from making the discovery that the boy made. The puppet will never become aware of itself as a visible body and, therefore, will not lose its bodily grace. The boy, however, did recognize himself in the mirror; he recognized the Spinario's beauty as his own, which resulted in one jubilant moment; from that moment on, however, he was on a downhill path. Like Mr C, nothing is left for him but to mirror himself in the ideal but unattainable grace of, for example, the marionette.

von Kleist's story about the boy in front of the mirror shows remarkable similarities with Lacan's psychoanalytical account of the development of a sense of self in what he calls the mirror stage. According to Lacan, it is at this moment that the ego comes into existence as a mental refraction of that image. Lacan founds this concept of the ego on Freud's remark that 'The ego is first and foremost a bodily ego: it is not merely a surface entity, but is itself the projection of a surface' (Freud, 1961b, p. 19). With this statement, Freud acknowledges that our experience of self is always circumscribed by and derived from the body. An important difference, however, between the Freudian and the Lacanian account is the role granted to vision in the inauguration of the self. In the Lacanian mirror stage, vision is given a hitherto unimagined prominence. Through the mirror stage vision, body and self are inextricably intertwined. It is through the mirror stage that one enters the scopic domain. Lacan calls it the 'threshold of the visible world' (Lacan, 1977, p. 3). Furthermore, what is formed in the mirror stage will be the rootstock of later identifications.

The moment the child recognizes itself in the mirror image is, according to Lacan, 'an identification in the full sense that analysis gives to the term', namely 'the transformation that takes place in the subject as he assumes an image' (Lacan, 1977, p. 2). It is a jubilant moment, for the visual imago of the body is more coherent than the organic disturbance and discord of the body as felt. However, as Lacan points out, the jubilant moment of assuming an image is in fact a mis-recognition: It is an identification with something outside the self. Like the marionette, the mirror presents an image of a more ideal body, but unlike Mr C, the child in front of the mirror does not see a more ideal other body. Instead, it recognizes the image as its own, and identifies with it as the form – or Gestalt – of the ego in a way that conceals its own lack. This conflation of self and other on the threshold of the visible world will remain an important ingredient of vision and self-awareness for the rest of the subject's life.

Inner mimicry

Lacan presented his idea on the mirror stage at a conference of the International Psychoanalytic Association in Marienbad in 1936.[2] Three years later, John Martin published his *Introduction to the Dance*. Like von Kleist, Martin believes that looking at dancing bodies is so attractive because – in the ideal case – it allows for direct contact with the moving force – or vis motrix – behind the movements seen. This way, dance can compensate for something lost, for a lack. 'All art, with the dance in the forefront', Martin argues, 'is a matter of compensation. It deals not with what we already have, but with what we lack' (Martin, 1939, p. 130).

For Martin then, as for von Kleist, this lack results from the loss of a more natural or original state.

> The oneness of all dance lies in the fact that in its every manifestation it consists of movement arranged in form to provide compensation for suppressions and unfulfillments in life experience. (Martin, 1939, p. 132)

According to Martin, modern life does not offer the opportunity to live one's inherent potentialities to the maximum. Therefore, a set of circumstances must be set up to compensate for the denials and suppressions that occur in daily life. Dance is such a circumstance: it helps to restore the individual to what Martin calls 'normal and harmonious functioning' which, according to him, is lost during the process of ongoing civilization. This re-creational function of dance applies not only to dancing, but also to looking at dance. Looking is what Martin does.

For Martin, the vis motrix or soul of the dance does not reside in the centre of gravity of the moving body, but in its ability to tap into pan-human emotions. Through looking at dancing bodies, we have, according to him, access to a universal core of basic human feelings made evident through the choreography and performance, and this is accomplished through what Martin calls the body's capacity for *inner mimicry*. Through inner mimicry, we 'cease to be mere spectators and become participants in the movement that is presented to us and though to all outward appearances we shall be sitting quietly in our chairs we shall nevertheless be dancing synthetically with all our musculature' (Martin, 1939, p. 3). Through a reaction of bodily responsiveness not unlike the one of the man photographed by Renée Kool in front of Courbet's *L'origine du monde* (Figure 8, see Chapter 6), we become aware of how it feels to make movements seen without actually executing them. This bodily awareness allows us to feel not only movements, but also feelings of the body seen executing them because, according to Martin, there is a natural connection between seeing movement and feeling it, and between feeling movement and emotions. In this way, dance provides us with a universal language that unites all mankind.

Martin writes not about marionettes, but about modern dance. With his explanation, he provided a rationale for the (then) new modern dance. His theory revises understandings of the meaning of bodies seen on stage and no longer explains them in terms of Swans, Willis or other ghosts in the representation of a narrative on stage, nor reducing them to sexualized appearances constructed to satisfy the desires of a male gaze. Nevertheless, his explanation also presents problems. As Susan Foster points out, Martin presupposes an intrinsic and mechanistic connection between seeing, movement and feeling.

> [He] imagines movement to be the transparent vehicle of an innermost, and hence, pan-human emotional realm. In the same way that muscular action intrinsically links to emotion, so for Martin, the individual psyche replicates the tensile patterns of the universal human condition. (Foster, unpublished paper)

Martin builds his universalist claim on the presupposition that the body spontaneously maps the contours of the psyche, the veracity of its pronouncements a direct product of its intrinsic connection to interiority. This would suggests that all human bodies have the capacity to feel what other bodies feel just by looking at them. Yet, notwithstanding the claimed universality of this pan-human realm of feelings, not all humans appear to have equal access to it, nor do all human bodies appear to be equally able to express these universal emotions through their movements. Martin's body willingly mimics some bodies, but refuses to mimic others. For, while his concept of inner mimicry serves to justify Martha Graham's impersonations of native American dances and Helen Tamiris's embodiment of the plight of the Negro, he also uses it to criticize black and native dancers for being too specific to represent the universal. Like von Kleist, he blames this difference on disorders perceived in the bodies seen, claiming that it is their racial and ethnic features that obstruct direct access to their vis motrix. They have to leave behind their racial and ethnic features in order to enter a realm Martha Graham and Helen Tamiris possess 'by nature'.

Spacing-out the mirror stage

In Martin's model there are two bodies: the one seeing and feeling in the auditorium, and the one seen as a spectacle on stage. The gap separating them is bridged by an instantaneous mapping of one body onto the other within the act of looking. Although Martin does not refer to psychoanalysis, it is hard to miss the analogy of his theory of seeing dance with the Lacanian scenario of the mirror stage. And, surprisingly, Silverman's critique of the Lacanian mirror stage in *The Threshold of the Visible World* (1996) parallels Foster's critique of Martin on at least one crucial point, namely the presupposed instantaneous and natural character of the jump from the body felt to the body seen needed to support the claim for universality.

In Lacan, this instantaneous (mis)recognition serves the constitution of the self. In Martin, it allows for a direct contact with the feelings of the other body seen on stage. In both cases, the automatic alignment of the visual and the corporeal as presupposed by the authors produces a mixing up of what is self and what is other. This mixing up is the core of Silverman's critical engagement with Lacan. Her project is not to undo this mixing up, but to 'space-out' the instant in which it is supposed to take place, in order to show it as a process of interaction that takes place over time and in which different positions are involved. A process, furthermore, that is not the result of natural or intrinsic mechanisms, but is mediated by culture.

Silverman turns to the work of Henri Wallon and Paul Schilder, two contemporaries of Lacan, for a slightly different account of the mirror stage, one focusing less exclusively on vision in the constitution of the self.[3] According to Lacan, the mirror image itself – the image of the body seen in the mirror – is sufficient to induce the 'assuming of an image' that inaugurates the self. Silverman comes up with an alternative model, in which the bodily ego has a sensational as well as a visual dimension. A model, furthermore, in which the bodily ego is not unified but produced in an ongoing process of 'laborious stitching together of disparate parts' (Silverman, 1996, p. 17).

Following Wallon, Silverman calls these components the 'exteroceptive' ego and the 'proprioceptive' ego. The former is comparable to the mirror image; the latter refers to 'the egoic component to which signifiers of deixis like "here", "there", and "my" are keyed' (Silverman, 1996, p. 16). Silverman points out the etymological roots of the term proprioceptivity as a combination of proprius – 'which includes among its central meanings, "personal", "individual", "characteristic", and "belonging"' – and capere, 'which means "to grasp", "to conceive", and "to catch"' (ibid.) She concludes that proprioception signifies 'something like the apprehension on the part of the subject of his or her "ownness"' (ibid.). Proprioceptivity is bound up with the body's sensation of occupying a point in space, and with the terms under which it does so. It encompasses the muscular system in its totality, and involves a non-visual mapping of the bodily form on the basis of a gathering together of otherwise disparate and scattered sensations, provided by the various sense organs. And, finally, it 'provides something which the specular imago alone could never provide – something which Wallon elsewhere, in an unfortunate choice of words, designates as "presence"' (ibid.).

Embodied presence

Silverman calls Wallon's use of the term 'presence' unfortunate, because it does not correspond with the idea of presence as being non-relational and independent from cultural interference. The concept of proprioceptivity as Silverman develops it on the basis of Wallon, includes all the effects of physical interactions not only with the physical environment, but also with other bodies. These interactions take place 'within culture'. Through these interactions the subject comes to have a body that is sensationally marked by gender, race and sexual preference. It is in relation to this sensationally marked part of the bodily ego that we perceive things as being exterior to us, that the specular image might be said to be 'outside'. It is also this relationship that, in Silverman's re-reading of Lacan through Wallon, allows for the jubilant experience of 'here and now-ness', in which the image outside and the bodily self as experienced from the inside seem to merge into one. Silverman writes:

> As I have already suggested, the visual imago cannot by itself induce in the subject that *meconnaisance* about which Lacan writes. The experience which each of us at times has of being 'ourselves' – the triumph of what I have been designating the *moi* part of the bodily ego – depends on the smooth integration of the visual imago with the proprioceptive or sensational ego. When the former seems unified with the latter, the subject experiences that mode of 'altogetherness' generally synonymous with 'presence'. When these two bodies come apart, that 'presence' is lost. (Silverman, 1996, p. 17; italics in the original)

'Presence' thus understood is not something given and observed to be 'over there,' but an effect produced as a result of a particular relationship between visual imago and proprioceptive ego, one that suggests its own absence in a jubilant moment of (mis)recognition. 'Presence' results from a relationship between a body seeing and the image of a body seen, and both sides of this relationship as well as what connects them, are embedded in culture. Within this relationship, 'presence' is an experience of confirmation of the body seeing rather than a quality observed or present in a body seen. It is for these reasons that Silverman calls the term 'presence' unfortunate. It is for these very same reasons, however, that I think such a relational notion of 'presence' might be a useful concept for rethinking the intense effects produced by the bodies seen on the contemporary stage.

Wallon's notion of 'presence' might help to turn presence into a critical and productive concept again, because it allows for a relocation of 'presence' to the eye of the beholder. It helps to understand presence as an effect of culturally mediated vision, constituted within a relationship of seer and seen. Furthermore, Silverman's re-reading of the mirror stage through Wallon makes it possible to theorize how the eye of the beholder is embodied, that is, how the beholder, as a body, is involved in how this beholder relates to bodies seen on stage. Finally, this new notion of theatrical presence might contribute to a re-reading of John Martin's concept of inner mimicry as a way of making things one's own through a process of non-visual mapping of what is seen on a culturally inflected body.

Yet, if the mirror stage is to be used as a model for the theatre, some questions have to be answered. Is it possible to use the mirror image model of identification to describe the relationship with bodies other than one's own? Can the model of mirror stage identification describe what happens when bodies see bodies that are not comparable to a mirror image in the strict sense, as for example John Martin's 'assuming' of the image of Martha Graham's body? This immediately brings up a second question: Why does his proprioceptive ego smoothly integrate with Martha Graham's white female body, but not with the image of a black dancer? What forms the limits of mirror stage identification?

The screen

In Lacan's account of the mirror stage, the subject 'assumes an image' and as a result a transformation takes place, one that inaugurates the self as we know it. The image assumed differs from the body as felt in some important respects. Lacan puts it as follows:

> The fact is that the total form of the body by which the subject anticipates in a mirage the maturation of his power is given to him only as

> a Gestalt, that is to say, in an exteriority in which this form is certainly more constituent than constituted, but in which it appears to him above all in a contrasting size (*un relief de stature*) that fixes it and in a symmetry that inverts it, in contrast with the turbulent movements that the subject feels are animating him. (Lacan, 1977, p. 2; italics in the original)

The difference between the body as experienced from within and the image perceived in the mirror, is of crucial importance for the assuming of the image to take place. It is because the image is more attractive than the body as felt, that the infant subject is prompted to recognize it as its own. At the same time, Lacan's reliance on the mirror – that is, on a tableau in which the visual image seems to be a direct extension of the physical body of the child and naturally related to it – implicitly but strongly suggests the importance of correspondence. As a model, the mirror suggests that the child recognizes itself in the image, because it is an image of its *own* body and is recognized as such because it is the same. At this point as well, Silverman argues for more 'space' in the Lacanian model. She sets out to theorize the possibility of a disjunctive relationship between visual imago and sensational body, a relationship that is mediated by culture. She proposes to replace the mirror with Lacan's concept of the screen from *Seminar XI*.

In *Seminar XI*, Lacan repeats what he had said previously, namely that for his or her visual identity the subject relies on external representation. In his explanation here, however, he does not refer to a mirror, but to what he calls the screen. The screen is not reflective like a mirror surface, but is opaque. The images of the screen appear as a result not of mechanical reflection, but of cultural intervention. The screen makes visible what culture admits, and blocks out the rest. In *Male Subjectivity at the Margins* (1992), Silverman proposes imagining the screen as the repertoire of representations by means of which our culture figures all of those many varieties of 'difference' through which social identity is inscribed (Silverman, 1992, p. 150).

von Kleist's story of the boy in front of the mirror offers an example of the working of the screen. In this story, the boy does not recognize himself in a mere image of his body in the mirror. His jubilant moment of (mis)recognition takes place at the moment that his mirror image resembles a classical statue that he and the narrator had admired shortly before. The classical statue represents an ideal of bodily beauty, especially at the time von Kleist wrote this text. 'Copies are familiar and to be found in most German collections' he says (von Kleist, 1982, p. 215).

The statue represents an ideal that belongs to the cultural repertoire of images that makes up the screen.

There is no existential connection between the screen image and the subject who is defined through it, and no necessary analogy. The screen can invite identification with images that are rather different from a mirror image in the literal sense. This distance, however, does not imply freedom. To be able to successfully invite identification, the image must possess a certain ideality. It must appeal to a desire, just as the mirror image appeals to the child who identifies with it. What can appear as ideal, is culturally mediated. Furthermore, what is culturally mediated is not only what can appear as an ideal image, but also who is allowed to identify with it. It is not enough that the subject (mis)recognizes him or herself within an image. In the Lacanian model of the field of vision, the (mis)recognition can only be successful if the subject is apprehended in that guise by the other. The alignment of the proprioceptive with the exterioceptive ego involves more than the look of the body seeing and feeling. It involves the look of others, the look of the other. Identification, it turns out, is a three-way rather than a two-way transaction (Silverman, 1996, p. 18).

The experience of being seen has a tremendous effect on our sense of self, and on how we experience the relation between self and world. It can confirm our self(mis)recognition, but it can also deny positive identification, as happens to the boy in von Kleist's story. The older man, being the authority, has introduced the boy to the culture represented by the Spinario, and to the notion of bodily ideality represented by it. He taught him how to see it and, more important, taught him to see 'through' it, and even to see himself 'through' it. Yet in the end, what is decisive is not how the boy sees himself or would like to see himself, but how he is seen. As soon as the boy recognizes himself in the mirror image, he turns to the man for confirmation of what he sees. The man, instead of providing such confirmation, destroys the self-image of the boy, leaving him in confusion. This negative possibility is implied by Lacan's account of the mirror stage, but he does not theorize it. This possibility, however, is central to Silverman's elaboration on the Lacanian model.

Silverman refers to Franz Fanon's *Black Skin White Masks* (1986) in order to illustrate the devastating effects of such a denial on the construction of a self-image. Fanon – a black man born in one of France's former colonies, and 'raised on a steady diet of Gaelic culture' – regards himself as 'French' rather than 'black' (Silverman, 1996, p. 27). At least, as long as he remains in the colony. When he moves to France, however, he is suddenly confronted with what he looks like in the eyes of those who are not black. He describes how he feels himself being addressed by the look of others in a very disconcerting way. He feels himself being seen through images of blackness and in this way, forced to identify with an imago he did not identify with before.[4]

von Kleist's and Fanon's stories demonstrate how we become aware of ourselves as bodies not only through reflections in mirrors, but also through the look of others seeing us. Through our awareness of being seen, we become aware of ourselves as a spectacle for others and as part of the spectacle of the world. This is not the result of the look of individual people projecting things onto one another. Fanon states how it is impossible for him to indicate the source from which he is seen through these images of blackness. Whenever he tries to point at where it comes from, the source seems to evaporate. His awareness of being seen results from an unlocalizable mechanism that makes him feel 'photographed' in an undesirable way. This is what Lacan theorizes as the mechanism of the *gaze*.

Silverman describes the Lacanian gaze as something that impresses itself upon us through the sensation each of us at times has of being held within the field of vision, of being given over to specularity (Silverman, 1996, p. 167). The gaze is everywhere and nowhere at the same time, and manifests itself more through its effects than through its source. It has much in common with the Sartrean concept of *le regard* in his famous theoretical story of the voyeur peeping through a keyhole, which has become deeply familiar in contemporary critical theory.[5] Although Lacan engages himself with Sartre's text only once, his account of the gaze, according to Silverman, 'begs to be read in tandem with that text' (Silverman, 1996, p. 167). This is not only because of the similarities, but also because reading them in tandem can help to clarify the differences between them. Since these differences are crucial to the critique of vision as presented in *The Path of the Dancer's Soul*, I will follow Silverman's argument and start with a brief look at her reading of the Sartrean model.

The gaze

Sartre distinguishes between two different acts of looking, and illustrates the difference between them with the example of a voyeur peeping though a keyhole, like the seer in Duchamp's *Etant donnés* discussed in the previous chapter. The voyeur is absorbed in the spectacle in front of him to such an extent that his own embodied presence escapes his attention. In his condition as voyeur, he is devoid of self-consciousness and paradoxically, this is synonymous with a certain transcendence. But then, suddenly, something happens that reminds him of his 'being there', for example, the sound of footsteps behind him. The sound conjures him out of his state of 'nothingness' into existence. He becomes aware that he himself is part of the visual world as well, seen from the point of view of the other evoked by the footsteps. From having a pure consciousness of things, the voyeur now becomes aware of himself as a spectacle, and it is through this awareness that a consciousness of self is produced in him. This sense of self is inextricably bound up with an awareness of his own specularity, of his being visible to others, to the other.

The voyeur suffers his specularity as the loss of transcendence. His transcendent position is transcended, and he feels himself no longer master of the field of vision. His mastery is unmasked as an illusion, as he suddenly realizes that he himself is also 'on view'. Like von Kleist's Mr C before him, Sartre relates this falling into a state of self-awareness with the Fall. To function as a spectacle, and thus to exist for the other, is to lapse into a 'fallen' state (Silverman, 1996, p. 165). Silverman points out how Sartre associates this experience of specularity with a whole series of psychic 'symptoms', all of which are somehow indicative of the condition of being in a relation of exteriority to one's self. These symptoms are then imaginarily converted into a list of antithetical values to produce le regard (Silverman, 1996, p. 164). She demonstrates how these oppositions contribute to a central opposition at work at the heart of Sartre's model of the field of vision: that of subject and object or, to be more precise, the antithesis of a pure or absolute subject and an object that is its absolute opposite.

As Silverman observes, Sartre characterizes this 'pure subject' in terms that are so in excess of human capacities that it ultimately becomes a phantom category, one that cannot be associated with an actual pair of eyes. She therefore calls this absolute subject a 'specular agency' fantasized into existence on the basis of the voyeur's apprehension of his emplacement within the field of vision. It is more a subjective effect than the result of an actual situation (Silverman, 1996, p. 166). Yet, at the same time, Sartre attributes human (or at least anthropomorphically inflected) functions to this agent, as a result of which the absolute subject or other appears as an imaginary rival.

The Sartrean model of the field of vision informed what is now a strongly established line of feminist criticism that gained its fullest theoretical articulation in film studies. Theorizing the role of the camera in the construction of what is to be seen on film, how the camera mediates in the appearance of female bodies, and how this construction invites particular ways of looking at them, has assisted in unmasking the appearance of bodies on screen as product of subjective vision, constructed to satisfy what is called 'the male gaze'. What might otherwise pass for a mechanical inscription of what bodies look like, thus becomes exposed as a construction implying a specific point of view.

These developments in film theory in their turn proved to be productive for the theatre as a tool for analysing the appearance of women in the scenarios on stage. However, the transposition of this model to the theatre has its problems. As Sue Ellen Case (1995) observes:

> The power relations of the Gaze in narrative cinema seemed homologous to operations of spectatorship in the theatre. Yet the obvious but astonishing fact that this construction of the Gaze helped to inscribe a role for the technology of the camera and the screen in the critical understanding of gender politics somehow eluded our attention. We neglected to comprehend fully how the power relations in the visual were necessarily cojoined with a mechanical apparatus for seeing. (Case, 1995, p. 330)

Theatre does make use of techniques to direct attention and to guide the look of the audience, such as various modes of focalization, yet it lacks a technique of vision comparable to the camera. Furthermore, although some theatre traditions and theatre practices do exploit the voyeuristic pleasure of peeping through a keyhole into another world while remaining invisible in the dark, a wide range of other theatrical practices demonstrate that this is not a necessary characteristic of the theatre event, nor is it a necessary precondition for the intense experience referred to as 'presence'. On the contrary, in many contemporary theatre practices, it is precisely the direct and explicit relationship with the audience that contributes to the intensity of the theatre as a 'live' experience, directly present and visible over there.

These bodies seen on stage are not objectified the way certain female bodies on film are turned into spectacle and criticized in feminist film theory. They do not necessarily capitalize on voyeuristic pleasures. Often they are explicitly theatrical. They look back at the audience, showing that they know that they show. They present a challenge to the audience to make a distinction between the act of showing and what is actually there to be seen, in this way challenging the old anti-theatrical prejudice reflected in Mr C's rejection of the self-conscious behavior of living dancers as displaying 'baroque mannerism'. That is, they challenge the use of theatricality as a pejorative term that refers to false or inauthentic behavior, so concerned with being seen that appearance takes precedence over what is shown. What is more, they challenge the presuppositions that underlie this negative understanding of 'theatricality', namely that it would be possible to see things 'as they are', and that this 'givenness' becomes distorted in a self-conscious act of showing. In this way, these 'exhibitions' of bodies on stage point to some of the limitations of the Sartrean model for an analysis of vision in the theatre.

In Sartre's model, the point of view of the voyeur is transcended and exposed in its objectivity. The model thus tends to confirm rather than to deconstruct the opposition of 'mere' representation and a true presence behind it. According to Silverman, Sartre misses the crucial opportunity implicit in the concept of the 'looked at look' – that is, the opportunity to theorize it in relation not to objectivity, but to subjectivity. She acknowledges the value of his insistence upon the eye's embodiment and specularity for feminist attempts 'to divest the male look of its false claim to be the gaze', yet also points out that what happens this way is not so much a deconstruction as a relocation of the notion of a transcendental eye (Silverman, 1996, p. 166). von Kleist's story demonstrates what such a relocation involves, and how it helps restore a stable sense of self through the construction of a new, all-seeing point of view of the absolute subject that, like Evelyn Fox Keller's paradoxical scientific subject (see Chapter 3), has dissociated him or herself from any specific location.

Paradise regained?

In von Kleist's third story, Sartre's imaginary rival materializes into an actual body and literally becomes a rival in a fencing match. In this story, Mr C recalls a visit he once paid to the estate of a Livonian nobleman. Mr C, along with being a very good dancer, is also a virtuoso fencer and is more than a match for all but one of his opponents. After he has defeated the sons of the nobleman, they challenge him to test his skills on a bear that their father is having raised on the estate. Much to his own surprise, Mr C appears to be unable to beat the bear. The bear makes him feel highly uncomfortable. What appears to be most formidable is the way the bear looks at him and seems to be able to read his soul.

> The earnestness of the bear was robbing me of my composure, thrusts and feints followed on one another, I was dripping with sweat: in vain! It was not merely that the bear, like the world's leading fencer,

> parried every one of my thrusts, but to my feints he reacted not at all (a feat that no fencer anywhere could match). Eye to eye, as though he could read my very soul, he stood with his paw poised for the strike, and if my thrusts were not in earnest he simply did not move. (von Kleist, 1982, p. 216)

Mr C feels threatened by the bear, while in fact it is he who attacks the bear. The bear only defends itself. What is really threatening Mr C is the bear's look. This unexpected look undermines Mr C's notion of himself not only as a master fencer, but also as master of the field of vision. Like the Sartrean voyeur, he is robbed of his sense of transcendence and mastery, and feels exposed as part of the visible world.

In von Kleist's story, Mr C finally beats the bear in the conclusion in which he puts the three stories in a broader, cosmological perspective. In this conclusion, he himself produces a new point of view that allows him to transcend the threatening encounter with the bear, and to regain his sense of self as master of the field of vision. According to him, the three stories demonstrate how

> in the organic world, as reflection grows darker and weaker, grace emerges ever more radiant and supreme. But just as two intersecting lines, converging on one side of a point, reappear on the other after their passage through infinity, and just as our image as we approach a concave mirror, vanishes to infinity only to reappear before our very eyes, so will grace, having likewise traversed the infinite, return to us once more, and so appear most purely in that bodily form that has either no consciousness at all or an infinite one, which is to say, either in the puppet or in a God. (von Kleist, 1982, p. 216)

The conclusion exploits what Paul de Man describes as 'the idea of innocence recovered at the far side ..., of paradise regained after the fall into consciousness ..., of a teleological and apocalyptic history of consciousness'. This is a history in which the gap separating being and consciousness will finally be closed again in a state of immediate presence, a return to paradise lost, to a state of primordial wholeness. As de Man observes, these powerful and seductive ideas of the romantic period make it easy to forget how little this pseudo-conclusion in fact has to do with the rest of the text (de Man, 1984, pp. 267–268). The conclusion suggests a symmetry that is not supported by the stories it claims to be based upon. For, although it is not difficult to link the conclusion to the first story about the marionette owing its superior grace to its lack of consciousness, the question remains as to how to consider the other possibility for the return of grace – God – a bodily form with infinite consciousness. None of the three stories provides a clue about the relationship between infinite consciousness and bodily grace. On the contrary, again and again they stress the first part of Mr C's conclusion that grace and reflection are to be understood as antagonistic. The second story – about the boy in front of the mirror – is an illustration of how self-consciousness hinders bodily grace. When the boy self-consciously tries to assume the image of the Spinario, he is not able to repeat the graceful attitude he unconsciously assumed a moment ago. No clue is given as to how to overcome this troubling situation again. The third story about the bear, is even more puzzling in relation to the teleological and symmetrical history suggested in the conclusion. Where to position the bear in the teleological history of consciousness 'from puppet to God'?

In the first and the second stories, Mr C and the narrator are the subjects observing grace – or the lack of it – in other bodies. They explain this lack of grace in terms of a relationship between consciousness and grace in the bodies observed. In the third story, however, Mr C is the object of the sharp eye of the bear. Here, so it would seem, it is the bear's lack of consciousness that allows him not to perform, but to read movements in a superior way. The bear sees through Mr C's attempts to fool him, and this gives Mr C the feeling of being exposed.

The look of the bear undermines not only Mr C's self-image, but also the Sartrean symmetrical model of the field of vision. Although the story of the bear is similar to the Sartrean scenario in some ways, it differs from it in important respects. For example, unlike the voyeur, Mr C can hardly have been unaware that he was the object of the bear's vision, otherwise it would make no sense to fence with the bear. Therefore, it cannot be his sudden realization that he is seen by the bear that bothers him and undermines his sense of self as master of the situation. It is rather, the way in which he feels exposed to others, just as it is others – and not the bear threatening him – that he has to convince of the truth of his conclusion, of his version of 'how it is' with the world. It is the narrator's confirmation that Mr C seeks with this conclusion, and it is through the involvement of this third party that he sets out to regain his sense of self and of mastery. Mr C's story about the bear thus indicates an asymmetry at work in the field of vision that Sartre's model does not account for and is better theorized through Lacan.

Lacan stresses the asymmetry of the look with which the voyeur peers through the keyhole, and the gaze with which he is surprised. The

Lacanian gaze is not the same as the look of a subject: the gaze escapes the category of the subject. At one point, Lacan defines it as 'the presence of others as such', but is eager to make clear that it is a 'function of seeingness' and is not to be mistaken for the look of concrete other subjects (Lacan, 1981, p. 84). This 'function' precedes any individual act of looking, and is that out of which the look somehow emerges much as language might be said to pre-exist the subject and provide him or her with signifying resources. For this reason, Silverman calls the gaze 'the manifestation of the symbolic within the field of vision' (Silverman, 1996, p. 168). The Lacanian gaze disrupts the symmetrical opposition of subject and object, of self and other, of body and mirror image. It is neither subjective nor objective. The gaze appears as a third term mediating in the constitution of what is self and what is other. It points to the interference of culture in the experience of 'just looking', and in how we perceive the visual field.

It is the gaze that can be seen at work in John Martin's body mimicking the body of Martha Graham, perceiving the ideal image presented by her white, female body as 'naturalness', while rejecting the bodies of black and native American dancers as distortions of the natural essences he perceives in Graham. It is not only because Martin is white, like Martha Graham, that her body can appear to him as more natural and truthful, but also because within the culture in which Martin's observations take place, whiteness signifies this complex intertwining of the 'natural' and the ideal. It is therefore not only because Martin himself is not black or native American, that seeing native or black bodies does not produce that feeling of recognition, the feeling of being able to spontaneously feel the feelings of the bodies seen. As Silverman's example of Fanon demonstrates, recognition is not only based on sameness or similarities; it also involves a certain culturally specific ideality, as a result of which – within a particular cultural and historical context – some bodies can appear as 'how it is', that is, as the manifestation of a 'naturalness' that other bodies lack.

Gestures of exposure

In their staging of von Kleist's text, Rijnders and De Châtel foreground the asymmetry at work in the field of vision on stage through a most effective, if not terribly spectacular, mise en scène. They present Von Kleist's text as an act of exposure, literally an exhibition of a body on stage, and a very specific one at that. Their staging resembles the gesture of showing which, according to Mieke Bal (1996a), is typical of a museum. It is a gesture that points to things, saying 'look, this is how it is'. Although Bal deals with exhibitions, and not with theatre performances, her analysis of the exhibition as an act of exposure is useful for an analysis of the exhibition of bodies on stage.

In an act of exposure, something is made public and, in Bal's view, this involves making public the deepest held views and beliefs of a subject. Exposition, therefore, is always also an argument. In publicizing these views, the subject exposes himself or herself every bit as much as the object. 'Such exposure is an act of producing meaning, a performance' (Bal, 1996, p. 2). The performative character of exposure however, remains hidden in the typical constative gesture of exposure of the museum that asserts 'this is how it is'.

In order to lay bare the performative aspects involved in constative gestures of exposure, Bal proposes to conceive of gestures of showing as discursive acts analogous to speech-acts. In such acts, three positions or 'persons' are involved: The 'first person' – the exposer – who tells a 'second person' – the visitor – about a 'third person' – the object on display (Bal, 1996, pp. 3–4). In the Rijnders/De Châtel performance, the actor represents the first person. He stands on the left-hand side of the stage and recites von Kleist's text *On the Puppet Theatre*. He acts as though he is giving a lecture, addressing the audience that represents the second person. On the right-hand side of the stage, there is the dancer making simple dance movements. He does not speak; he is just there, acknowledging the presence of the audience through his look. He represents the third person, the object on display, the body exhibited.

The object is framed in a way that enables the statement to come across. This frame helps to read what is seen according to the logic implied within the frame. An important part of framing in this performance is the change of title, from von Kleist's *On the Puppet Theatre* to Rijnders's and De Châtel's *The Path of the Dancer's Soul*, which is a quote from the text. This new title turns the dancer into the object of the argument. This object seen does not participate in the conversation. The object is there to substantiate the statement. However, unlike the speech-acts that make up von Kleist's text, in the performance the object, although mute, is present. This is important, for it allows for a critical move.

When the speaker is near the end of the text, the dancer makes an unexpected move: he leaves his position on the right-hand side of the stage, walks towards the speaker, and looks at him. This move appears to be highly uncomfortable for the speaker, who tries to avoid the dancer's eyes, to ignore his presence near him, and to focus his attention on the audience. The speaker begins to repeat the entire text at double speed, as well as his demonstrative gestures.

The unexpected move by the dancer resembles that of the bear in Mr C's story, just as the response of the speaker resembles that of Mr C. Like Mr C, the speaker tries to counter the threat by attempting to impose his story – his vision of 'how it is' – on the audience, and to deny the disturbing look of the object of his vision. He literally ignores the dancer and repeats what he has already said, as though to stress that this is 'how it is'. However, the dancer, who is visually present before the eyes of audience, prevents him from accomplishing what Mr C manages to accomplish in the text. On stage, the speaker loses his authority, and his demonstration of 'how it is' with the object of his attention turns into an exposition of himself as an embodied subject.

The unexpected move by the dancer offers the audience a second point of view. This critical move presents an invitation to the audience to look in a different way. This moment presents an example of what Mieke Bal (1991) has called the 'navel' of the performance text (discussed in Chapter 5). The dancer turns from object seen into a subject of focalization who redirects the audience's attention towards the speaker. Of course, this speaker was visible from the outset and was well aware of his being visible. Yet, the look of the dancer exposes him in a different way.

The dancer's move causes an actual re-reading of the entire text by von Kleist in a different way exposing it not as a demonstration of truth, but as a nervous attempt by the speaker to keep going. By redirecting attention, the dancer turns the demonstration of 'how it is' with the dancer – with bodies seen – into a performance that instead exposes the observer as subject seeing. His critical move turns the demonstration of 'how it is' with the body of the dancer seen, into an exhibition of the speaker as body seeing. As a result, the speaker is exposed as a desiring subject, exposed to himself, and also to the 'presence of others as such'. In this process, the look of the audience stands in for the Lacanian gaze. This is not to say that the audience is gazing, nor that their look is the gaze: The Lacanian gaze refers to being conscious of one's visibility. *The Path of the Dancer's Soul* uses the theatre situation to externalize the mechanism and in this way to expose its functioning.

Exposing the audience

In *The Path of the Dancer's Soul*, the desiring subject exposed is a body on stage. Lacan's account of the field of vision however, allows for an analysis of how the look of the body seeing in the auditorium is also under cultural and psychological pressure. Here, moments of 'presence' prove to be particularly instructive. With a Lacanian model of the construction of the field of vision, it is possible to understand the feeling of intense closeness to bodies directly 'present' on stage, not in terms of the characteristics of the object, nor as the result of the immediateness of the presentation, but as the effect of a culturally mediated way of looking.

Even before we become conscious of having seen something, perception has been processed in all kinds of ways. The look is exhorted from many sides to perceive and affirm only what generally passes for 'reality.' In *Male Subjectivity at the Margins* (1992), Silverman proposes to replace 'reality' with dominant fiction. The 'fiction' in 'dominant fiction' foregrounds the constructed nature of what passes for reality. The 'dominant' points to the fact that there is more than one fiction possible and, also, that these different possibilities do not have equal access to the 'reality' status. Calling reality a fiction undermines its ontological claims on truth, and opens it up to change and cultural difference. Here again, however, it is important to notice that this does not imply unlimited freedom. The dominant fiction is not a story told by the self, nor can it be changed by an individual. It is the story within which both world and the self are produced, and it is at work at levels we do not consciously control.

The 'mirror stage moment' in von Kleist's text can be read as a demonstration of the dominant fiction at work. What is needed in order to expose this fiction at work, is to redirect attention analogous to the critical move performed by the dancer in *The Path of the Dancer's Soul*. von Kleist's narrator tells a story about a boy in front of a mirror. Yet, this story is not that of, nor that told by, the boy: it is the story of the older man observing the boy. It is as much about him as about the boy observed. Interesting in this respect is the account the narrator gives of his own reaction to what happens to the boy. Recalling the moment the boy pointed out that he looked like the Spinario, the man says:

> I indeed had noticed it too in the very same instant, but either to test the self-assurance of the grace with which he was endowed, or to challenge his vanity in a salutary way, I laughed and said he was seeing phantoms. (von Kleist, 1982, p. 215)

His denial of the boy's discovery is not the correction of a mistake. It is not an exposure of 'how it is', confronting the boy's subjective point of view with the objective truth seen from a transcendental point of view. On the contrary, it is a denial of what he, too, has seen, but does not or cannot believe. His response is a subjective reaction dictated by cultural values and psychic pressure, dictated perhaps by the same cultural presupposition that was already at stake in the first story, namely that a living and self-conscious being can never approach the beauty and grace of a marionette. Seen this way, the older man's reaction is an aggressive one towards someone who seems capable of approximating what he himself cannot. His reaction demonstrates how he translates a threat to his own ego into a disorder perceived in the boy.

In the same way, to expose the audience as a desiring subject, what is needed is an explanation of the appearance of bodies on stage as the product of a cultural and psychological determined look, rather than as simply being there to be seen. Here again, moments of 'presence' can provide a point of entrance for they present a kind of audience variant of the constative gesture of 'this is how it is'. 'Presence' – understood in terms of Silverman/Wallon as a successful stitching together of an image of a body seen and the proprioceptive ego of the body seeing – can function as a meaningful pointer, indicating that the body seen appeals to a desire of the body seeing. This desire is not a deformation of reality, but constitutive of what is perceived as reality. To demonstrate what such an exposure of the subject of vision could mean, I will end this chapter with a little thought experiment, and conceive of the mirror stage as described by Lacan as a staged situation like the Rijnders/De Châtel performance. In order to 'look back' at the subject of vision, I will perform a move analogous to the one proposed by the performance, and expand Bal's analysis of the act of exposure as a discursive gesture to this academic demonstration of 'how it is'.

Looking back at Lacan

Like von Kleist's story about the boy in front of the mirror, the Lacanian account of the mirror stage is a story about a child and not that of, or by, this child. It is the story of Lacan seeing and observing the child, and it is therefore, as much about the subject 'Lacan' as it is about the child. By 'the subject Lacan', I do not mean Jacques Lacan as a historical person, but the subject of vision as it is produced in his text. I wonder what his text about the mirror stage can tell me about this subject of vision as a body. What could it mean to understand Lacan's observations as the product of an embodied look under cultural pressure to perceive the world from a pre-assigned point of view? Especially the jubilant recognition is telling in this respect, provided that it is approached as proposed by the performance, that is, starting from the question of to whom this bodily image appears as more ideal, and what desire this image appeals to.

According to Lacan, it is the organic disturbance and discord of the body as experienced from the inside that prompts the child to assume whole body-image perceived in the mirror. Lacan acknowledges the possibility of moments of collapse of the alignment with the ideal image, moments that result in what he calls the fantasy of the body in bits and pieces. Silverman points out that this explanation in terms of disintegration and failure is only one way of apprehending the heterogeneity of the corporeal ego, and a very particular one, too. It is a way of apprehending that is inextricably tied to aspirations concerning wholeness and unity. Silverman proposes to reverse Lacan's argument, and argues that the opposite is actually true. It is not that the child seeks wholeness because it experiences itself as fragmented. According to Silverman 'it is the cultural premium placed on the notion of a coherent bodily ego that results in such a dystopic apprehension of corporeal multiplicity' (Silverman, 1996, p. 21). It is because of the cultural appreciation of bodily wholeness that the child experiences its bodily fragmentation as dystopic. Both the identification with the unitary image of a body in the mirror, and the fantasy of the body in bits and pieces, are the result of the cultural premium placed on unity that informs the Lacanian subject. Silverman likewise draws attention to the constructed nature of what can appear as an ideal image. Again, she points to the fact that the relation between the image the subject identifies with, and the body as felt, has to be understood as a product of culture, rather than naturally given. Yet at this point, Silverman's elaboration on Lacan seems to open onto a more profound critique as well. A further implication of her reading seems to be that the specific relationship of body image and body as felt, as observed by Lacan might be culturally informed as well. In Chapters 8 and 9, I further explore this relation between the Lacanian account of the formation of the 'I' in the mirror stage and the cultural context from which this 'story' emanates. I end this chapter with a first move in this direction.

Lacan characterizes the coherence to which the classical subject aspires as 'the armor of an always alienating identity' (Lacan, 1977, p. 4). It is the result of a cultural logic in which 'wholeness' – a coherent bodily ego – signifies psychic health. Susan Buck-Morss (1992) suggests that this cultural imperative has to be understood in the context of sensory alienation typical of modernity. According to her, the significance of

Lacan's theory emerges in the historical context of modernity as precisely the experience of the fragile body and the dangers of fragmentation. She presents her critique of Lacan as part of a re-reading of Walter Benjamin's famous essay 'The Work of Art in the Age of Mechanical Reproduction', which appeared in 1936, the year that Lacan first presented his ideas on the mirror stage.

Buck-Morss describes how, according to Benjamin, the very essence of the modern experience is shock. The technologically altered environment of modern city life, the factory, and modern warfare, expose the human sensorium to physical shocks that have their correspondence in psychic shock. It is against this cultural background of this kind of Western, big city modernity that the unitary, well outlined body-as-image in the mirror appears desirable. The mirror image is a body as seen from the point of view German writer Ernst Jünger describes as 'second consciousness'. Jünger calls the technological order dominating modern life a mirror that reflects back an image of the body that alters our awareness of ourselves as embodied beings.

> In the great mirror of technology, the image that returns is displaced, reflected onto a different plane, where one sees oneself as a physical body divorced from sensory vulnerability – a statistical body, the behavior of which can be calculated; a performing body, actions of which can be measured up against the norm; a virtual body, one can endure the shocks of modernity without pain. (Buck-Morss, 1992, p. 33)

In modern Western big city life, war causalities, as well as industrial and traffic accidents, have become accepted as a feature of existence, and have forced modern man to develop a second consciousness that is indicated in the ever-more sharply developed capacity to see oneself as an object. Jünger writes: 'It almost seems as if the human being possessed a striving to create a space in which pain ... can be regarded as an illusion' (Jünger quoted in Buck-Morss, 1992, p. 33).

The construction of a second consciousness that manifests itself as the result of identification with an image of the body, divorced from the sensory vulnerability of the body felt from the inside, reacts against the shocks that make up modern life. Benjamin – relying on the Freudian insight that consciousness is a shield protecting the organism against stimuli – comes up with a neurological explanation pointing in the same direction. To protect itself against the constant bombardment of shocks, the ego employs consciousness as a shield, blocking the openness of the synaesthetic system of the body, thereby isolating present consciousness from past memory. As a result, experience becomes impoverished.

> Being 'cheated out of experience' has become the general state, as the synaesthetic system is marshalled to parry technological stimuli in order to protect both the body from the trauma of accident and the psyche from the trauma of perceptual shock. As a result, the system reverses its role. Its goal is to *numb* the organism, to deaden the senses, to repress memory: the cognitive system of synaesthetics has become, rather, one of anaesthetics. (Buck-Morss, 1992, p. 16; italics in the original)

Buck-Morss suggests that the cultural imperative of assuming the image of the unitary body in the mirror is a symptom of the threat imposed on the subject as a vulnerable body under the pressure of modern life and a technologically altered environment. Furthermore, this cultural premium not only informs the subject to identify with a particular bodily image, but also the specific relationship of body image, and body as felt, that makes up the very foundations of the Lacanian model of ego formation. His conceptualization of subject formation resonates with the 'disembodied' character typical of the modern understanding of vision, as well as the vigorous privileging of vision over the other senses. It is through looking that modern individuals are understood to gain insight into themselves and the world, to such a extent that the 'I' of the looker and his or her eye almost become conflated. Elements of this fable can still be seen at work in today's understanding of ourselves as embodied subjects, and in how we experience our relationship to the world around us.

Buck-Morss's reading suggests the possibility of understanding the Lacanian model as the product of a particular historical moment, as a conceptualization that appeals to a desire at work within the dominant fiction of modernity. At the same time, her historicization of the Lacanian model invites one to think beyond Lacan, and to undo the simultaneity of over-stimulation and numbness that have become the characteristics of the synaesthetic organization of the modern subject. Such rethinking might also grant new actuality to Martin's concept of inner mimicry. Inner mimicry seems to offer the option of regarding not only the bodily image as constitutive of the embodied self but also, vice versa, the embodied self as constitutive of the image of a body perceived. This would imply a slightly different reading of proprioception as a combination of *proprius* and *capere*. That is, proprioception meaning to grasp, conceive or catch what is seen through a process of bodily responsiveness. Inner mimicry, then, would not mean feeling what the other body seen is feeling, but using one's own bodily feelings and kinesthetic responses to make sense of a body seen. Understood in this way, inner mimicry does not present a link to an original universality, but describes a bodily process of culturally specific meaning-making. It would come to mean a way of making things one's own through a process of non-visual mapping of what is seen on a culturally inflected body. Inner mimicry as the process of the mapping of a body seen onto the body seeing as a means to make sense of the body seen, might help to overcome the mono-directionality implied in the Lacanian model, and to call attention to the ways in which reactions of bodily responsiveness contribute to the construction of what is perceived as 'just looking'.

8
Death, Digitalization and Dys-Appearance

> [W]hen one knows how much the souls of beasts differ from our souls, one understands very much better the reasons that prove that our soul is of nature entirely independent of the body, and consequently that it isn't subjected to die with the body; then, inasmuch as one doesn't see other causes that would destroy it, one is naturally inclined to judge that it is immortal. (Descartes, 1977, p. 152)

In proving that the body is a mechanical, mathematical entity, free of all soul attributes, Descartes laid the groundwork for modern scientific medicine. He hoped to discover ways to prolong embodied life. However, since life cannot go on forever, he also felt the need to prove the immaterial nature of the rational soul, and thereby its immortality. Descartes' scheme serves to combat death on all fronts. His strategy for overcoming death is precisely to capture the body fully in the third person. It is the body of the other that Descartes anatomizes, his own body is then reconstructed on such a model.

I do not want to suggest that Descartes is to be held personally responsible for what has come to be known as the Cartesian paradigm, or that the mind – body dualism characteristic of this paradigm has to be understood as the product of his personal intentions. But I do think that observations like the one quoted above can serve as an indicator of the subjective perspective at work in visions of the world 'as it is' according to Cartesian world view. They can serve as an indicator as well of the kind of subjective desires and fears involved in the willing identification with this point of view and illuminate the relationships between this point of view and a particular conception of what bodies are and what they can do.

In the previous chapter, I introduced Silverman's (1996) reading of Lacan to explain identification with an image of a body seen as always already taking place under the pressure of a cultural gaze. Silverman's rereading turns Lacan's mirror stage model, from an explanation of the origin of subjectivity, into a model that describes how the seer as subject and the visible world come into being in relation to one another in an ongoing process and also how culturally specific images of bodies mediate in this process. 'Assuming', as Lacan puts it, the coherent and unitary image of the body seen in the mirror allows for a stabilization of the turmoil of the body experienced from the inside, while at the same time it presents the child with an identity separate from its surroundings. For the first time, the child is not absorbed into its environment, but becomes aware of itself as different from it. Thus, the mirror stage not only invites the child to identify with a body seen, it also mediates in how the child becomes constituted as a body seeing in relation to what is seen. This perspective is what is at stake in this chapter.

Lacan speaks of 'a jubilant assumption of the specular image', but also makes it clear that the stability of the unified body image, even in the so-called normal subject, is always precarious. Unity cannot be taken for granted as an accomplished fact, for it must be continually renewed.[1] Silverman expands on this ambiguity and offers a model that can account for how culturally specific images of bodies seen play a part in the ongoing life-long process of stabilizing the self. This way, she is able to point out the relation between the Lacanian account of the formation of the I in the mirror stage and the cultural context from which this 'story' emanates.

In this chapter, I will engage with this story in dialogue with a performance that promotes a return to the historical anatomy theatre for a reconsideraton of the body as the locus of our looking. This performance is *Holoman; Digital Cadaver* by Mike Tyler (1997/1998).[2] *Holoman; Digital Cadaver* draws attention to how, in both visualizations of the body produced by anatomy and medical science, and in the theatre, strategies of de-theatricalization are used to bring the seer closer to bodies seen. It also emphasizes how these strategies can be seen to alienate the seer from his or her body as the locus of looking. Doing so, *Holoman; Digital Cadaver* makes us aware of what might be called the 'other side' of Silverman's rereading of the Lacanian model. Not only is our awareness of self as body seen mediated by the repertoire of images that make up the visual unconscious, but also images of bodies seen mediate in a specific awareness of ourselves as (disembodied) seers of these bodies.

Anatomical theatre

The Renaissance anatomy theatre marks the emergence of a new knowledge about the body. Eventually, this knowledge would come to represent *the* knowledge of the body, while at the same time – as Jonathan Sawday points out – the anatomical dissection would come to represent the model of scientific investigation *par exellence*. The historical anatomy theatre symbolizes the emergence of a particular constellation of ideas and practices underlying what has become the dominant conception of the body, including the prevailing notions of how the body can be known and what it means to know. This popularity of anatomy, Sawday argues, cannot be understood solely from raising the ban on the formerly forbidden practice of dissection, nor simply as a result of the superior quality of the knowledge thus produced. Rather, the body of anatomy is part and parcel of the development of modern individualism, and of the modern scientific world view. Dissection turns the body into a mute corporeal object, separated from and opposed to the Cartesian disembodied eye/I as the site of subjectivity, thought and knowledge. Additionally, the 'culture of dissection' (Sawday, 1995) marks the beginnings of what Foucault has analysed as the 'surveillance' of the body within regimes of judgement and punishment, as well as an early crystallization of the modern Western sense of interiority.

In the anatomy theatre, dissections were not so much investigations as they were demonstrations of the authority of the anatomist. The anatomist demonstrated his knowledge to the audience. The audience was invited to take up the position of subject of vision as represented by the anatomist and to recognize what is shown as being the truth about the body and, therefore, about themselves as bodies. This perspective is clearly visible in the theatrical set up of the historical anatomy theatre, presenting a gesture of exposure not unlike the one performed in *The Path of the Dancer's Soul* discussed in the previous chapter. With Bal (1996a) we can understand this gesture as a constative gesture that involves three different positions. In the first place, there is the anatomist, who provides a 'first person' exposure of the corpse. The anatomist addresses the audience representing the second person involved in the gesture of exposure. The audience is invited to understand their own bodies through the parameters presented by the body seen over there. This body is the third person, literally subjected to the dissection performed on it, turned into a mute object, there to prove authority of anatomist.

Constative gestures are always also performative in the sense that they are acts of making meaning. In the anatomy theatres too, the body was not only demonstrated, but also performed. On the one hand there was the material body, cut open and falling apart, transformed by rigor mortis, lifeless, smelly, chaotic. On the other hand, as a kind of guide for the dissectionist, there was the anatomical text, usually a classical source: descriptive, organized, classificatory. The difference between them is according to Jonathan Sawday one of the central tensions in early modern anatomy.

In the Renaissance anatomy theatre, the space between the anatomical body and the body of anatomical knowledge was constantly negotiated. Sawday shows how this process of negotiation can be seen reflected in the development of a range of representational strategies that produce the body as machine body in the seventeenth century. He describes how, during the period which opened with the publication of Vesalius *De Humani Corporis Fabrica* in 1543 and continued roughly to 1640, scientists like Columbian explorers dotted their names, like place names on a map, over the terrain which they encountered (Sawday, 1995, p. 23). Natural philosophers of the early seventeenth century conceived of the body in terms of territory, a yet undiscovered country that was to be explored and charted in an anatomical atlas. This process was truly colonial, in that it appeared to reproduce the stages of discovery and exploitation which were, at that moment, taking place within the European encounter with the new world. First the body was mapped, named and the new terrain observed. The second phase was to harness this new terrain for the use of the discoverer:

> Intrinsic to such a project was the creation of the body's interior as a form of property. Like property, the body's bounds needed to be fixed, its dimensions properly measured, its resources charted. Its 'new' owner – which would eventually become the thinking process of the Cartesian cogito – had to know what it was that was owned before use could be made of it. (Sawday, 1995, p. 26)

The reduction of the body to a mechanical entity, 'owned' and controlled by an immaterial cogito, did perhaps not begin with Descartes, but the Cartesian thinking that suggests that the operations of the body have to be analysed in terms of automata or moving machines can, according to Sawday, serve as a summation of half a century of voyages into the interior of which Descartes was the heir:

> Mechanism offered the prospect of a radically reconstituted body. Forged into a working machine, the mechanical body appeared

> fundamentally different from the geographic body whose contours expressed a static landscape without dynamic interconnection. More than that, the body as a machine, as a clock, as an automation, was understood as having no intellect of its own. Instead it silently operated according to the laws of mechanics. ... As a machine, the body became objectified: a focus of intense curiosity, but entirely divorced from the world of the speaking and thinking subject. The division between Cartesian subject and corporeal object, between an 'I' that thinks and an 'it' in which we reside, had become absolute. (Sawday, 1995, p. 29)

This inaugural moment of the invention of the anatomical body was highly theatrical in character, and occurred in a highly theatrical space. During the centuries that followed, this theatricality disappeared from view as strategies of de-theatricalization and semiotic repositioning (see Fox Keller discussed in Chapter 3) began to produce the objective body of medical science as we know it today.

A convenient cadaver

In *Holoman; Digital Cadaver,* Mike Tyler returns to the anatomy theatre for a confrontation of state of the art, objectified, anatomical visualizations of the human body with the body of an actor representing the human body that was used to produce them. His protagonist is a fictional character whose life and death parallel that of real-life murderer J. P. Jernigan. Jernigan was executed in Texas in 1993, but not before donating his body to science, or more precisely, to the Visible Human Project.[3] The goal of this project initiated by the National Library of Medicine (USA) is to create an information resource which will make possible highly detailed 3D navigations of what is called 'a representative male and female cadaver'.[4] The researchers working on the Visible Human Project speak of a 'renewable cadaver, a standarized patient, and a basis for digital populations of the future. Not only can we dissect it, we can put it back together again and start all over' (Spitzer and Whitlock, quoted in Thacker, 2001). To become such a convenient cadaver, Jernigan's body underwent MRI (Magnetic Resonance Interferometrie) and CT (Computerized Tomography) scanning. Subsequently, it was frozen to minus 70° Celsius and cut into 1mm-thick slices. Digital photographs of each of the 1,871 slices, together with the scans – a total of 15 gigabytes of electronic information – were processed in a computer to produce a digital resurrection of the cadaver, a ghost on the Net representing 'universal human meat' (Tyler, 1997/1998).[5]

J. P. Jernigan was a white man. The representatives of the Visible Human Project selected a white man to represent the universal human meat. Although his makers did plan right from the start to 'do' more bodies, and have started to do so,[6] the choice for a white man to be the first can hardly be called incidental. It is a choice, furthermore, that, taking into consideration the average population of death row, testifies to a certain determination. Tyler in his performance commented on this choice in a way that was literally spectacular. Actor Frank Sheppard, an African American man, played Holoman in 'white face', or, in this case it is better to say, in 'white body'. Prior to each show, his entire body was painted white with 'glow in the dark' make up, turning him into what seemed to be the uncanny re-appearance of the repressed other in the guise of the same. And Holoman presented an uncanny reappearance in more ways than this.

The mise en scène of *Holoman; Digital Cadaver* recalled the set up of the historical anatomy theatre combined with the secrecy and cleanness of the modern morgue. The audience, sitting on a stand, looked down on a bare stage. There, they could see a simple metal stretcher on wheels and next to the stretcher a big white projection screen. On the stretcher, the outline of a human body was covered with a white sheet. A voice-over reminded the audience that cadaver dissection has been illegal for centuries, forbidden by the church. The audience was reminded of how curious anatomists had to work in secrecy and steal bodies from the gallows. Eventually however, during the Renaissance, dissection became legalized and turned into a public spectacle attracting much attention. Special theatres were built in many European cities. The bodies used in public dissections were usually the cadavers of convicted criminals so that the spectacle of the dissected body could also function as an instrument for moral education. Dissection could be imposed as an extra punishment in addition to the death penalty; it was considered a humiliation and it meant the denial of a last resting place. The horrible fate of the criminals on the dissection table served as a frightening example of what misbehaviour could lead to. Inscriptions on the walls of the former anatomical theatre in de Waag in Amsterdam still remind one of this function.[7] Given this then, Jernigan's digitalization continues an old tradition of disciplining the living body through the dead.[8]

But more is at stake. The anatomy theatre has its place in a theatrical economy in which the body is marked as the Other and encapsulated in a structure designed to neutralize its supposed threat. Public dissection then was part and parcel of the development of the Cartesian paradigm in which the body is marked as something we have instead of something we are; a body, furthermore, that we look at rather than look from.

Carried away on a data stream

Descartes' scheme turns the human cadaver, torn apart and transformed into a body of knowledge, into a metaphor for the Enlightenment project, and Visible Humans into its personifications. This is not just because of Descartes' fearful contempt for bodily matters. Nor is it only because anatomy as an 'opening up' in order to see deeper or hidden parts is deep in the heart of the Enlightenment project. It is also because, ironically, the destroyed and disappeared body, repressed by the Cartesian enterprise, can be seen to return in the metaphors of science itself. At the heart of the scientific project of the Enlightenment, Barbara Maria Stafford (1991) observes a remarkable set of body metaphors. The sensual body, rejected as being senseless, returns in the images the mind uses to make sense of knowledge, to literally embody abstract ideas in order to grasp them and to sense their implications. In the Visible Human Project this knowledge literally takes the shape of a human body.

With the Visible Human Project, man created a body of knowledge after his own image, or so it seems. A body that is literally enlightened, consisting of sheer energy. But this mirror image appears to be marked by lack. Struggling with the gap between his disembodied psyche and the senses that connect his body to the world, the protagonist of *Holoman; Digital Cadaver* points to some of the controversies and dilemmas that haunt the Cartesian subject disconnected from his objectified body. In *Holoman; Digital Cadaver* the mute object of scientific examination rises from the dead and talks back. While 'his' digital body is downloaded from the Internet, he speaks to us as a disembodied psyche, trapped within the computer. He tells us how he was persuaded to give his body to science. His confessions testify to how this noble deed was motivated by fears not unlike the ones that drove Descartes to the dissection table.

The representatives of the Visible Human Project took advantage of Holoman's fear of death and finiteness, primal fears that must become especially poignant on death row. Unlike his predecessors in the historical anatomical theatre, Holoman never thought of his dissection as an extra punishment or as a denial of a last resting place. To him, it sounded like an opportunity to escape the finitude of his body and to remain forever present. Holoman grew up in a modern Western society typified by a certain 'disembodied' lifestyle. Protected from direct corporeal engagement with the outer world in many ways, it is easy to forget the material basis of our existence. Technologies of communication and transportation allow us to transcend what used to be the natural limits imposed by the body. Now that we are becoming increasingly familiar with images of people, with artificial persons existing as bytes and bits of optical and aural messages, the distinction between physical and digital reality seems to be fading. Painfully aware of the limitations of his material existence, Holoman is extra sensitive to this suggestive aura of the digital media. To him, digitalization appears as a survival technique, a way to transcend the limitations of his vulnerable and mortal body. This survival technique seems to offer him something similar to what religion previously provided.

At first sight, Holoman's strategy to overcome death may seem to be the opposite of the one put forward by Descartes. Descartes distances 'himself' (where 'himself can be understood as essentially a thinking and seeing entity) from his body as mere matter. As a result, he is able to grasp the body through the metaphor of the machine. Or, the other way round, his conception of the body in terms of a mechanical entity, presents him with a position for the subject of vision at a certain distance from the body seen as such. It allows him a perspective on the body rather than a perspective from the body.

Holoman, on the other hand, identifies with his own dead body. He identifies with a digitally produced image of his own corpse, an image that he does not seem to perceive as an image at all but, instead, as a digitally produced continuation of his physical body. He identifies with this digitally produced image and begins to think that the image will secure his eternal presence. Like the child in the Lacanian mirror stage, Holoman identifies with an image of his body seen from a point of view outside of it. To him, the digital image of his body is literally a means of stabilizing and counteracting the disturbing awareness of his sensitive and sensible body. Poor Holoman. His mirror image appears to be marked by lack indeed.

Come to your senses!

With his performance, Tyler shows the identification with an image of a dead body to be not the opposite of the Cartesian paradigm but to be its ultimate implication. His performance suggests that the Lacanian mirror stage based on identification with the image of another body seen, might be symptomatic of the de-corporealized Cartesian subject and the effect of internalization of the perspective on the body presented by the anatomy theatre. Descartes' scheme invites us to take up the position of the anatomist (from where the body appears as other) and to look at the body from there. But he also invites us to understand our own body as similar to the body seen over there. That is, he invites us to identify ourselves as bodies with the body seen. This position presented to us, therefore, is complicated and taking it up involves both a conflation of the image of the body presented by this knowledge with one's own body, and the taking up of a position from where this body is seen as other.

In the historical anatomy theatre, this perspective is clearly visible. The explicit gesture of exposure performed by the anatomical theatre sets up a relationship between body seeing (the audience) and body seen (the body dissected/the corpse) mediated by the subjective point of view represented by the anatomist. The de-theatricalized visualizations of the Visible Human Project on the other hand, suggest the absence of such a subjective perspective. The explicit theatricality of the historical anatomical theatre is replaced by a seemingly objective showing 'how it is'. This suggestive aura of the images, presented to him by the new media, make Holoman forget the distance between himself and the digitally produced images of his body. No wonder he decides to put his faith in digitalization as a man-made afterlife.

During the performance, however, Holoman comes to his senses again, and in more ways than one. First, he comes to his senses in the usual figurative sense of coming out of unconsciousness or folly as he wakes up and begins to sense the implications of his 'transubstantiation'. Second, it is also possible to read the expression 'Come to your senses' in a more literal sense, as an invitation or perhaps even an order to come towards the place where his senses are. Or, in his case, used to be. For, in his case, identification with a perspective on his body rather than from his body, has literally resulted in the erasure of his body as the locus of his looking. With Holoman's entrance into cyberspace, his body has ceased to exist as a place or locus. His transformation into a body of knowledge reduced his body to zero. Its fate is not unlike that of an archeological site, where the quest for invisible origins results in a complete destruction of the location itself. Holoman's body is reduced to bits and pieces, the pieces are brushed aside and the bits are used to produce a digital image.

During the performance, Holoman comes back to what he left behind, which is his sensible and sensing body. This suggests that he first must have left this body and somehow stepped outside of it. Usually, experiences of being outside the body are considered to be paranormal phenomena. Drew Leder (1990), however, argues for the opposite. A careful rereading of Merleau-Ponty's phenomenology leads Leder to conclude that the experience of being outside one's body is in fact a normal condition. Although in one sense the body is the most abiding and inescapable presence in our lives, it is also essentially characterized by absence. My own body is rarely the thematic object of my experience. My attention is directed intentionally towards the outside world and rarely dwells on my own embodiment. According to Leder it is precisely this phenomenological condition of our bodies that plays a crucial role in encouraging and supporting Cartesian mind–body dualism. He therefore proposes an analysis of the experience of bodily absence as a tool for understanding concepts of self.

The body, Leder observes, tends to disappear from our attention as long as it functions unproblematically. Largely taken for granted as the place of life, well-being and growth, the body is most manifest at times of dysfunction. At those moments, we experience the body as the very absence of a desired or ordinary state. The body appears as a force that stands opposed to the self. This leads Leder to propose a new concept as a tool for investigating the tension between the disappearing body and its troubling reappearance. He introduces the term dys-appearance to denote a mode through which the body appears to explicit awareness.

In dys-appearance, the prefix dys evokes several levels of meaning. In Greek, dys signifies 'bad', 'hard' or 'ill', a sense of meaning preserved in such English words as 'dysfunction', as well as in many terms for illnesses, such as 'dysentery', and 'dyslexia'. Dys can also be understood as a variant, now somewhat archaic spelling of the Latin root dis, which originally had the meaning 'away', 'apart' or 'asunder'. Dys-appearance, therefore, effects an attentional reversal of disappearance. The words dys-appearance and disappearance have an antonymic significance, while at the same time the homonimity of the words is meant to suggest the deep relation between these modes (Leder, 1990, pp. 86–87).

A place of intertwining

In *Holoman; Digital Cadaver,* Mike Tyler uses the hypertext anatomical atlas on the Net for what seems to be a postmodern re-valuation of the old meaning of the aesthetic experience as that which is perceptive by feeling; a form of cognition achieved through taste, touch, hearing, seeing and smell – the whole corporeal sensorium. In waking the digital cadaver from its condition of anesthesia, Tyler's performance can be read as a response to Walter Benjamin's demand (discussed in the previous chapter) to undo the condition of sensory alienation that he sees threatening modern man, to restore the power of the human bodily senses. Benjamin explains this condition of sensory alienation as the effect of the pressure put on human beings by modern city life and factory work. With *Holoman; Digital Cadaver,* Tyler suggests a different way of reading instances of sensory alienation in modern times. Holoman's identification with an image of his own corpse points to the centrality of other bodies and the body as other in our notions of self and of subjectivity, and draws attention to how this goes at the cost of the absence or disappearance of the body as sensing and thinking being. Holoman's state of sensory deprivation is the direct result of his identification with an image of his body as other, as corpse (Figure 11). This identification has turned him into a disembodied psyche, alienated from his body as a sensing and sensible being.

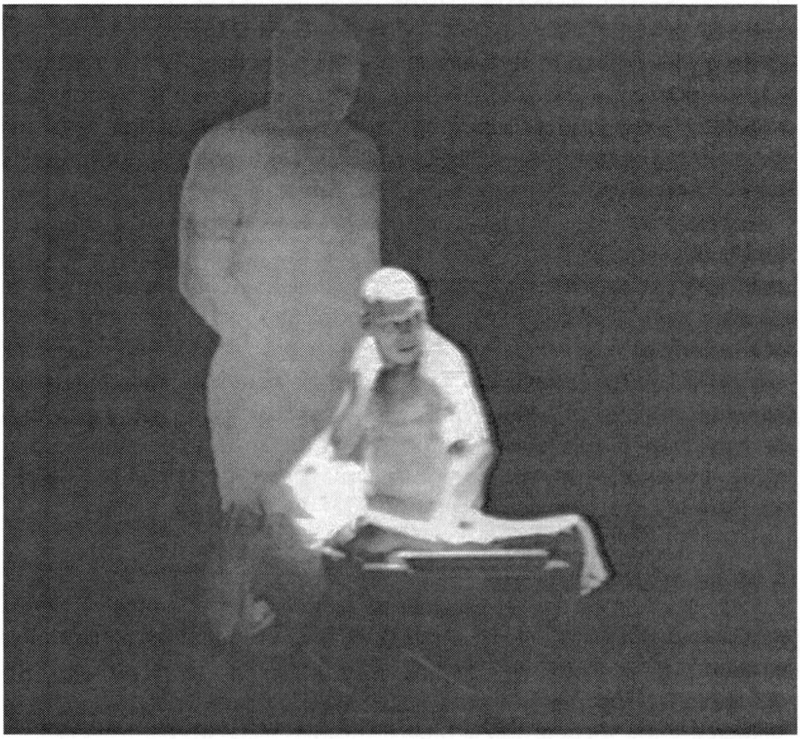

Figure 11 Photograph by Isabelle Jenniches showing Frank Sheppard in *Holoman; Digital Cadaver* by Mike Tyler. Reproduced with permission of the artists.

In *Holoman; Digital Cadaver*, the character acts as an internal focalizor who presents the seer in the theatre with a position to take up, a position from which the implications of this identification become apparent in a most compelling way. The audience is invited to assume the position presented by Holoman, who presents an invitation to the audience to step into his shoes and see it as if from there. Taking up this position turns the dys-appearance of the body as represented on stage into a sense of dys-placement, that is, it makes evident the usual absence of the body as location of sensory involvement with the world, precidely by making this body appear in unexpected ways. In Tyler's performance the digitally constructed dys-appearance is confronted with the physical body that disappeared in its making. In an attempt to make a direct connection with his digital *Doppelgänger*, actor Frank Sheppard looks straight into a digital camera placed in front of him. Then he quickly turns his head towards the projection screen behind him to look himself in the eyes. Before he can make a connection, however, the image turns its head as well and slowly recedes into the background, disappearing into the multidimensional hollows of Jernigan's digital body.

Holoman's condition recalls the psychotic condition called autoscopy. This is an extreme form of depersonalization in which the subject's ego is no longer centred in its own body. In this condition, the subject may see itself as if it were from the outside (Grosz, 1994, p. 43). *Holoman; Digital Cadaver* suggests this pathological condition might be not so much a strange aberration from some 'natural condition', as a consequence of a culturally produced self-awareness through images of bodies seen. The images of Sheppard playing Holoman, manipulated on the spot by computer artist Isabelle Jenniches, are like an electronic echo showing the paradoxical distance at the heart of the immediateness of the digital inscriptions (Figure 12). Holoman's digital resurrection is a mere reflection. He can only look at (t)his digitalized body; he can no longer look from it. It is exactly this intertwining of seeing and being seen that is of crucial importance to Merleau-Ponty's critique of the Cartesian mind–body opposition.

According to Merleau-Ponty, the presence of a body implies a particular reversibility. It is a 'place' that both sees and can be seen, that both touches and can be touched. Seeing and touching are each recorded on what Merleau-Ponty calls a map. These two maps are complete, but not super-posable. To paraphrase him, it is a marvel too little noticed that every movement of my eyes and every displacement of my body has its place in the same visible universe that I itemize and explore with them,

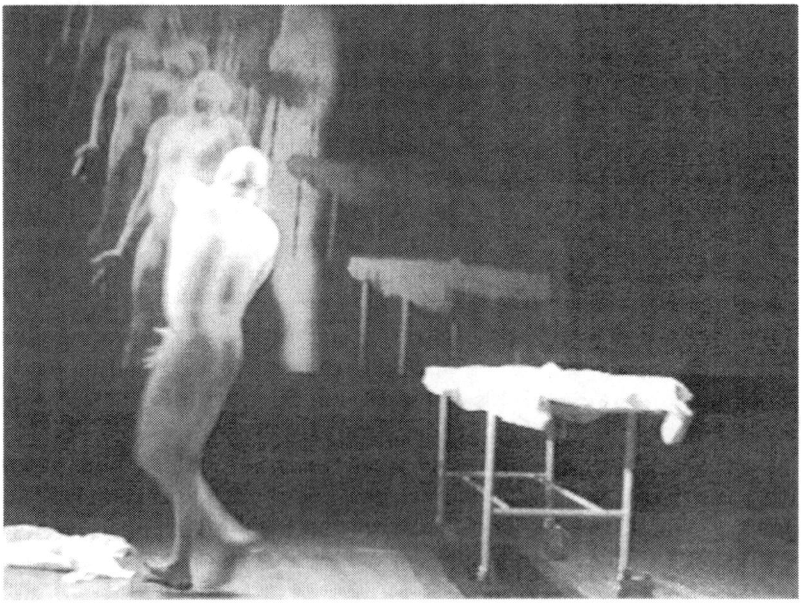

Figure 12 Photograph by Isabelle Jenniches showing Frank Sheppard in *Holoman: Digital Cadaver* by Mike Tyler. Reproduced with permission of the artists.

and that, conversely, every vision takes place somewhere in tactile space. This is what Merleau-Ponty calls a 'double and crossed situating of the visible in the tangible and of the tangible in the visible; the two maps are complete, and yet they do not merge into one' (Merleau-Ponty, 1968, p. 134). The tangible is not completely reducible to the visible, and vice versa. Nevertheless, we experience what we see and what we feel as referring to the same world. This is because it is the same body that both sees and touches. It is therefore not consciousness but the body that creates the unity of the world. Just as it is the difference between the two that can be used to make one aware of the synthesizing activities of the body and the limitations of unitary perspective based uniquely on vision. In *Holoman; Digital Cadaver* a computer screen prevents the connection between the visible and the tangible from taking place. When Holoman reaches out to touch his own body he feels the cold screen. But were there no screen, there would still be no spark generated between electronic image and tangible body. Instead, there would be a destructive flash terminating the actor as a living locus. It is only within the embodied look of the audience that the visible and the tangible body refer to the same world, yet in a way that disrupts the unity of this world.

Unlike the historical anatomy theatre, *Holoman; Digital Cadaver* does not confirm the truth of the Cartesian paradigm, but instead asks for the undoing of the modern transmutation. It presents an invitation to the audience as well to 'come to your senses', that is, to leave the Cartesian world of shadows and to study the double and crossed situation of the visible and the tangible (and the other senses as well, I would add) starting from a multi-sensuous engagement with an object that is not 'over there' behind the proscenium arch, but takes place in a situation of intertwining with a body over here, which is me.

9
Managing the Attention of the Audience

It is not consciousness but bodies that create the unity of the world. My body both sees and hears, smells and tastes, touches and feels, and creates the world as one world that is both visible and audible and tangible etc., and all at the same time. It is my body that creates the world as an object of perception. However, as Leder (discussed in the previous chapter) remarks, intentionally directed outwards to the world, my bodily investment in this world easily escapes my attention. My body is the blind spot in my experiences, the blind spot that is constitutive of a point of view or 'I' inside this world as it is laid out for me by my senses. The senses provide me with information about the world (including my own body), they are my connection to the world. Much more, they produce the world for me, including a sense of self in relation to what I perceive, and yet I tend to experience what I feel, see, hear, smell as *my* perceptions, that is, as perceptions of a 'me' as the origin of perception, a 'me' that uses the senses as 'window to the world' rather than being itself the product of sense perception. Like the blind spot in perspective, these enabling as well as constraining perceptual capacities of my body need to go unnoticed in order to experience myself as an 'I' amidst a world that is not the product of my acts of perception but their reason. At those moments that the investment of my body does become apparent, this investment tends to manifest itself as deception (my senses were misled) or imperfection (if only I could overcome the limitations imposed by my senses), suggesting that my senses need to be aided, corrected, disciplined to prevent deception from taking place, and thus protecting the idea of an otherwise objectively perceptible world. Jonathan Crary, in his *Suspensions of Perception: Attention, Spectacle and Modern Culture* (1999), presents a history of such practices of disciplining the senses in modern times through what he terms the 'management of attention' (see also Chapter 1).

Leder sets out to explain this condition of bodily absence in relation to the modes of absence inherent in the body. For, as he argues,

> Cultural variations are always played out upon the keyboard of possibilities presented by our corporeal structure. Only because the body has intrinsic tendencies toward self-concealment could such tendencies be exaggerated by linguistic and technological extensions. (Leder, 1990, p. 3)

In this chapter, I will take a slightly different approach and focus on what Leder refers to as 'cultural variations' and how such cultural variations manifest themselves in how we conceptualize our relationship to what we perceive, as well as how we conceive of ourselves as subjects of perception in relation to the world around us. This will bring me to a further historicization of the mirror stage model, again not in order to dismiss this model for being 'merely' historical but rather to demonstrate its usefulness for understanding how the world seen as stable and objectively given is the product of a body conditioned to perceive in culturally specific ways.

The Lacanian mirror stage offers an explanation of the coming into being of a sense of self as an 'I' amidst this world in which self-consciousness is understood as the consequence or effect of identification with an image of a body seen. Lacan's account of this moment is based on a distinction between the image seen in the mirror (that is presented as a more stable and therefore more attractive referent of the body) as opposed to sensory impressions felt from the inside of the body (which are presented as confusing and disturbing). His account of ego formation describes how these two get connected and how this then inaugurates the development of an 'I' that is the origin of perception. In presenting this as the normal development of the child, Lacan's story naturalizes this moment (including the initial separation of vision and the body) while at the same time their getting connected happens, in Lacan's words, 'in a mirage' (Lacan, 1977, p. 2), that is, it is something that is not self-evident and therefore needs explanation, and in this explanation the mirror plays a central part.

This miraculous moment is an important point in Silverman's (1996) critique of Lacan (see Chapter 7), in which she argues for more space in the Lacanian model (between a proprioceptive sense of self 'over here' and bodies seen 'over there') while she also brings in time by pointing out the historical and cultural specificity of the body image, as well as the fact that what Lacan describes in the mirror stage is not a single moment but the beginning of a life-long process. These elaborations bring her to a reading of Lacan's famous description of the mirror image being 'the threshold of the visible world' (Lacan, 1977, p. 3) as referring to the fact that this body image not only presents an appeal to identification with something outside the self, but at the same time presents the limit to all possible identifications. At this point, Silverman's elaboration on Lacan proves to be a most useful step beyond the subject-less and disembodied visions of the modern visual paradigm, and illuminates how we are always involved in what we see, even when seeing seems to be 'just looking'.

Yet, in at least one respect, Silverman's account remains remarkably faithful to the modern visual paradigm, and this is with regard to the opposition of the visual versus the body and the other senses. Actually, this opposition is one of the pillars of her argument, based as it is on an understanding of ego-formation as the laborious stitching together of a proprioceptive awareness of the body occupying a point in space and a visual imago. Silverman's model too implies an initial separation of the visual realm and the proprioceptive realm until the child learns to connect what it sees with what it feels.[1] Furthermore, in describing these two components as a proprioceptive ego that involvers a 'nonvisual mapping of the body's form' (Silverman, 1996, p. 16) and an exteroceptive ego which she equates with the image seen in the mirror (Silverman, 1996, p. 15), her model reconfirms the opposition of interiority versus exteriority characteristic of the modern subject, as well as the intimate connection between a private interior world as something first and foremost felt versus a public exterior world as something first and foremost seen.

What if we read 'threshold' not to mean some kind of limit or border that cannot be crossed, but rather (and perhaps at the same time) as the threshold that has to be crossed to 'step inside' the visible world? In this process, the mirror image, like an actor in the theatre, provides the child with a point of view or position to take up and 'step inside' the 'stage' or scene. This involves a process of connecting the experience of one's own body to the image of a body seen in the mirror – a process in which the child learns to align his/her perceptual experiences with this body seen. That is, to 'step inside' the mirror involves to adopt a mode of looking that places a premium on visual information as the point of reference from where to 'place' other perceptual input (or ignore such input).

It is as if I am describing a ballet class where dancers, always facing the mirror, learn to perceive their bodies first and foremost via the mirror image and ignore other perceptual information (including pain) that might distract them from what they see, all of this to become, in their turn, mirror images to an audience, perfect bodies in control, bodies that do not sweat and make no sound; bodies that can be perceived as mere spectacle.

This, again, makes the mirror such a meaningful choice to explain the constitution of this particular type of self-awareness. This is not only because of the ambiguity discussed in Chapter 7 (that the mirror image presents an image of a body that is both self and other) but also because the opposition of a body in the mirror and a body of flesh and blood *seems* to support the opposition of a body merely seen and a body felt. Seems, since this opposition is based on the equation of the mirror image as a visual phenomenon with seeing a (mirror) image as being a matter of visual perception only. This is the illusion constitutive of visual essentialism typical of the modern visual paradigm. Read this way, the mirror stage describes the moment that the body anticipates 'in a mirage' not only the culturally specific sense of self mediated by the mirror image, but also, and at the same time, the mode of perception that constitutes the world as a domain of pure visibility. That is, in order to make the discovery Lacan ascribes to it, the child has to anticipate a culturally specific mode of perception: the mode of perception that according to Crary is the product or effect of a history of disciplining the senses by means of culturally specific practices of managing attention.

Bentham's theatre

With this history, Crary points to what might be called the other side to the psychoanalytical account of the relationship between visuality and subjectivity in modernity. Whereas psychoanalysis represents what Elisabeth Grosz (1994) terms the 'inside out' approach to subjectivity, Crary refers to Nietzsche, Benjamin and Foucault and their approach of subjectivity 'from the outside in' to explain visuality (understood as a separate category isolated from the other senses) as a product of cultural practices of disciplining bodies. Visuality understood as a separate category, Crary argues, is the effect or consequence of such disciplinary practices that produce the seer as a disembodied I/eye in relation to an objectively visible world. The image par excellence of such disciplinary practices is, of course, Bentham's Panopticon, this famous example of Foucault that has such a central position in his *Discipline and Punish. The Birth of the Prison* (1995).

The Panopticon is a building. It is a prison built in a circle with a tower in the centre. The tower is pierced with wide windows that open onto the inner side of the ring. The circular building is divided into cells, each of which extends the whole width of the building. Each cell has two windows, one on the inside, corresponding to the windows of the tower, the other, on the outside, allowing the light to cross the cell from one end to the other.

> By the effect of backlighting, one can observe from the tower, standing out against the light, the small captive shadows in the periphery. They are like so many cages, so many small *theatres*, in which each actor is alone, perfectly individualized and constantly visible. (Foucault, 1995, p. 200; my italics)

The Panopticon is designed to induce a particular sense of self-awareness in the inmate. The position of the guard is only important as literally that, an imaginary position to be taken up by the inmate so that he or she could imagine him or herself to be seen from this subject position. In the Panopticon the staging of this 'see/being seen dyad' (Foucault) serves to direct the intentional awareness of the inmate towards his or her own presence as being seen. The inmate is forced into the position of the object of vision. Yet, as Foucault points out, its effect derives from the inmate inscribing him or herself in the power relation in which he or she simultaneously plays both roles. That is, the Panopticon derives its effect from the interiorization of the particular relationship between subject and object of vision of which the building presents a spatial metaphor. In the Panopticon, a disembodied all-seeing subject of vision is opposed to a body as object; subject and object are strictly separated. The disembodied subject is master over the body seen. This power is closely related to knowledge. The body/object cannot escape being known and mastered in its entirety by the subject seeing. It is this seeing, dissociated from the body and the other senses, that provides the subject with its power and knowledge. In short, the Panopticon presents a spatial metaphor of Cartesian subjectivity and modern notions of objective and disembodied vision related to it.

According to Crary, the modern notion of attention is a sign of reconfigurations of those disciplinary mechanisms:

> If disciplinary society was originally constituted around procedures through which the body was literally confined, physically isolated and regimented, or set in place at work, Foucault makes clear that these were but the first relatively crude experiments in an ongoing process of perfecting and refining such mechanisms. By the early

> twentieth century, the attentive subject is part of an internalization of disciplinary imperatives in which individuals are made more directly responsible for their own efficient or profitable utilization within various social arrangements. (Crary, 1999, p. 73)

No longer does the power of disciplinary institutions manifest itself by means of a guard or a prison wall producing the subject through enforcing discipline on a body quite literally through incarceration. Power now can be seen at work in practices of self-discipline to which the body is subjected. Power manifests itself in the ways in which responsibility is assumed for the constraints of power, and make them play upon oneself. At this moment, strategies of managing attention appear as means of sustaining, reinforcing or further developing the self-regulatory practices that produce the self. In this sense, managing attention is inseparable from the operations of Foucault's 'disciplinary' institutions, but as an inversion of his panoptic model in which the subject is an object of attention and surveillance. This inversion turns Bentham's Panopticon from a prison into a theatre.

Managing attention

Crary's analysis too turns the theatre into the paradigmatic example of a wide variety of cultural practices all aiming at maintaining and preserving the modern fable of vision as true and objective, of the possibility of seeing it 'as it is'. His history of managing attention in modern culture illuminates how cultural practices of disciplining bodies not only affect how bodies appear as objects of vision but also how they become subjects of vision; how bodies, subjected to a variety of disciplinary practices, learn to pay attention and perceive in culturally and historically specific ways. Attention emerges as a discursive and practical object right at the moment that vision and hearing had become progressively severed from the various codes and practices that had invested them with a level of certainty, dependability and naturalness. At this moment, strategies of managing attention appear as a promise of the possibility of transcending these selfsame subjective limitations and therefore as a means to safeguard the possibility of objective and true vision against the threat of subjectivity.

The theatre presents a model for understanding these practices as ways of organizing the relationship between the one seeing and what is seen, while at the same time theatricality (and by extension the theatre itself) is what has to be repressed in order to safeguard the illusion of the seen as evidence, as truth and fact; of the apparent autonomy of distance and separateness to the spectator; and of the visual aspects of the world seen as being properties of this world itself. Sometimes, this happens by means of devices and strategies that are literally 'Panopticon-like', for example in a conventional theatre situation where the audience is sitting in the dark, like the guard in the Panopticon, watching people on a well lit stage exposed to their look. This way, the theatre stages the relation between seer and seen, yet in a way that obscures precisely the relationality involved.

The theatre does not discipline its audiences; it does not turn people into disembodied subjects 'just looking'. Rather, it presents its audience with an address that resonates with the implications of already internalized modes of looking. Or, to speak with Freedman (Chapter 1), the theatre responds to culturally and historically specific spectator-consciousness. Either by supporting and reconfirming the expectations, desires and presuppositions as they are part of the spectator's mode of looking, or not. Important to this second possibility is a shift in direction of what Foucault terms the 'see/being seen dyad' between the Panopticon and the theatre.

The theatre is not simply a reversal of the situation in the Panopticon but its inverse. In the Panopticon, the separation of vision from the body and the other senses is instrumental in developing a new kind of self-awareness in which the inmate's visibility turns into proof of 'how it is' with him or her. The Panopticon suggestively separates this being visible from being audible and tangible. This does not mean that the inmate is unable to hear or feel. Rather, the Panopticon presents a model of the inauguration of a sense of self in which the experience of being seen represents a greater truth, or objectivity, to which the inmate has to subject itself. This greater truth manifests itself in a construction that captures the inmate in the field of vision of an invisible seer looking at the inmate from a position the inmate cannot see.[2] In the theatre, on the other hand, the particular 'staging' of the 'see/being seen dyad' does not serve the development of self-awareness of those on stage as bodies being seen, nor is the intentional awareness of the audience directed towards their presence as bodies seeing. Rather, this time, the spatial set-up serves to direct intentional awareness of the audience away from its own presence as body seeing and towards what is presented on stage. This happens by means of an inverse of the situation in the Panopticon, and this inverse responds to and reconfirms modes of looking typical of subjects who have already internalized the mode of looking (including the type of self-awareness related to it) of which the Panopticon presents a spatial metaphor. The inversion of the 'see/being seen dyad' turns the theatre into a staged version of the mirror stage in which the audience is invited to 'step inside' and identify-at-a-distance (Silverman) with bodies seen on stage.

This analogy between the theatre and the mirror stage illuminates the intimate relationship between seers towards bodies seen in the theatre and the psychological mechanism of mirror stage identification, while at the same time, as I demonstrate, the relationship between the theatre and the Panopticon allows for a further denaturalization of the separation of vision and the body implied within the psychoanalytical account of what happens in the mirror stage. In the following, I will perform such a denaturalization in dialogue with a theatre performance that confronts its audience literally with a mirror/stage. This performance is titled *Double Track*, created by choreographer Beppie Blankert, and based on Samuel Beckett's *Text for Nothing no.7*. This will also bring me to a revaluation of Martin's concept of inner mimicry (introduced in Chapter 7) to describe the involvement of the body in what we think we see.

Mirror/stage

In Beckett's *Text for Nothing no.7*, a narrator describes himself sitting on a bench in a small railway station, waiting for nothing. Trains and time pass by while his mind is absent, floating elsewhere, performing what Hugh Kenner has called 'fantasies of non-being' (1973, p. 119). The narrating subject in *Text for Nothing no.7* inhabits, as Jonathan Boulter puts it, a curiously 'located' space: 'a space in which the subject is at an ontological or at least psychological distance from itself' (1999, p. 7). Beckett's text confronts the reader with a merely disembodied voice speaking of itself to itself. In his text, this disembodied condition is related to particular type of vision 'as if from Sirius', and opposed to this 'heap of flesh' sitting over here in the railway station (Beckett, 1974, p. 37). Separation from the heap of flesh allows the mind to travel elsewhere while in fact 'I had not stirred hand or foot from the third class waiting-room of the South-Eastern Railway Terminus' (Ibid.). He wonders:

> [Is] that me still waiting, sitting up stiff and straight on the edge of the seat, knowing the dangers of laisser-aller, hands on thighs, ticket between finger and thumb, in that great room dim with the platform gloom as dispensed by the quarter glass self-closing door, locked up in those shadows, it's there, it's me. (Beckett, 1974, p. 38)

Beckett's description of the rather uncomfortable confrontation with this stiff body sitting straight on the edge of the seat reads like an experience I had, a long time ago, during the première of Hans van Manen's seminal creation *Live* (1979).[3] Prior to this duet between a ballerina and a cameraman, this cameraman used his video camera to film a single row of audience members on the balcony. The images were projected in real time on a large screen on stage. I still remember the strange mixture of curiosity and uncomfortable feelings when I began to realize that the people I saw there on screen where actually seated some places left to me. A few moments later, I too was confronted with an image of myself, sitting up stiff and straight, the programme book between my fingers, in that great room dim with the gloom of the house lights.

The gesture of the cameraman, apart from functioning as a sign of the liveness of the images that we would later on see projected on the screen, pointed to the ambiguous relationship between bodies seeing and bodies seen in the theatre. This gesture exposed my bodily presence in the audience as being part of the theatrical event while at the same time it made me aware of this bodily presence as something that, when made explicit to consciousness, prevents the 'stepping inside' to take place that would allow me to travel 'elsewhere, I don't know where'. Exposed by the camera as a body sitting there 'stiff and straight on the edge of the seat', I experienced a moment of bodily dys-appearance that made sensible the difference between bodily involvement in what is seen and bodily self-awareness. This difference, I will argue, is crucial to rethinking the relationship as well as the differences between the subject that says 'I' and the body as the condition of possibility of perception.

In 1986 and again in 1999, choreographer Beppie Blankert presented a staging of Beckett's text in a performance titled *Double Track*.[4] Playing with a huge mirror, Blankert turns the question of the location of the subject into a riddle for the audience. The audience finds itself located in between two raised stages. On the stage in front, a mirror shows the reflection of a dancer. This dancer is actually present on the stage behind the audience, sitting on a bench, waiting, like the character in Beckett's story. Just as Beckett's text presents its reader with self-reflexive remarks about the condition of the speaker sitting, watching and waiting in 'that great room' while his mind is floating elsewhere 'I don't know where', so Blankert's performance confronts the audience with self-reflection by exposing the audience as body seeing within the architecture of the theatrical event.

In Blankert's performance, Beckett's text 'migrates' from the domain of text-based theatre to dance. With this move, she sheds new light on questions of travelling, tracking movement, and positioning as they are brought up by Beckett in his text. Train travel turns into a metaphor for navigating attention through a performance, and the aesthetic logic of this performance into a positioning system defining where and how we, the audience, are; how this involves deducing one's position as seer from the address presented by the theatrical event; and how this address involves both positioning and displacement. *Double Track* does not present the seer with a position that suggests his or her bodily absence from the viewing scene. On the contrary, the seer is placed literally in the centre of the theatrical event and presented with an address that redirects attention from the world projected over there behind the shiny glass screen towards the 'heap of flesh' that is the locus of looking.

The spatial set-up is reminiscent of Plato's parable of man in the cave watching mere reflections of what is actually present behind their heads. But unlike the people stuck in the cave and deceived by mere shadows on the wall, the spectator of *Double Track* knows what he or she sees to be a reflection. When spectators enter the theatre room, both the mirror in front and the stage behind the auditorium are fully lit and a dancer can be seen present on the stage behind, sitting on the bench. The mirror splits the presence of the dancer in two. Two video monitors on the stage in front of the audience take the splitting to another level. They show the very same dancer sitting on a similar bench at a small railway station, waiting for nothing. Then, this dancer splits in two. The dancer, wearing a light coloured costume, is sitting on the bench and remains seated, while at the same time, another dancer in a dark costume rises from where the man in white is seated, and walks away.

This second person, as it appears, was actually seated on a similar bench behind the mirror, invisible to the audience. Invisible, until the light behind the mirror goes up. Through the see-through mirror the two dancers perform a duet that actually takes place in two different locations. They are on two different stages – one in front and one behind the audience. The dancer in front is seen through the mirror, while the one behind is seen reflected in the mirror. Often, it is hard to say who is where. The dancers constantly change positions: sometimes they are together behind the mirror, sometimes they are behind the audience, sometimes they are together on the same stage, sometimes they are not. This presents the audience with an address that positions them differently each time, while positioning them differently in visual and in auditory space. When the dancer seen in front starts to talk, his voice is heard from behind. When the two dancers perform a duet, the sound of their feet – coming from two different directions – undercuts the unity seen in the mirror.

The complicated address presented by the image seen in the mirror in front and the sound heard from different directions causes disturbing instances of bodily dys-appearance that redirect attention from the bodies seen in the mirror that is the stage, to the 'heap of flesh' perceiving them. *Double Track* undermines instead of supports a position for the seer as a disembodied and detached observer. The sounds deny the seer what he or she is granted by a more conventional theatre situation. That is, they prevent the seer from being absorbed into this world seen 'over there' and instead, make evident the body as locus of looking, seeing it as such 'over here'. These disturbing moments of bodily dys-appearance constitute what has to be suppressed or ignored in order for the spectator to leave the 'heap of flesh' in the auditorium and become absorbed in the image seen in the mirror. Yet, these moments can also function as what I have termed the navel of the performance (see Chapter 5), and shift attention from what is seen to the relationship between seer and seen. Such moments can show how this relationship involves a story.

His father's voice

The curious thing in experiencing *Double Track* is that although the sound confirms ones position at the centre of the event, and thus confirms what one can see (namely that one finds oneself at the centre of a performance taking place around oneself), it is nevertheless the sound that causes the uncomfortable sense of dys-placement. Although the audience of *Double Track* is reminded repeatedly of the fact that the unity perceived in the mirror is an illusion, it is difficult to perceive it as such. Although I knew from the start that what I saw in front of me was a mirror image of someone present behind me, I perceived the performance as if it were taking place in front of me. I continued to do so even though the sound of the voice of the person I saw speaking in front of me and the sound of his feet reminded me time and again that at least part of what I was seeing was, in fact, taking place behind my head.

Blankert's strategy of evoking dys-placement in *Double Track* recalls an observation made by Henri Wallon in his study *Les origins du caractère chez l'enfant* (1949). In this study, cited many times by Lacan, Wallon describes a situation of an infant smiling in recognition of its father's image in the mirror.[5] When the father speaks to the child, the child, surprised that the voice emanates from a different place than the image, seems shocked and turns from the image towards the father holding him. Wallon takes this as an indicator of the fact that the child has not yet grasped the differences and connections between its father's physical presence and his specular reflection. In normal development, the child will eventually learn to understand how the image in the mirror and the father's physical presence are different yet related to one another. In Lacan's mirror stage story, this 'being the same yet different,' is crucial for the pleasure that the child finds in recognizing its own image in the mirror, and for the promise this mirror image seems to offer. This is a promise of a unity and coherence that it does not feel from inside.

Wallon explains the child's confusion from its not (yet) knowing the difference and connections between mirror image and reality. This difference is something to be learned, after which it would cease to be a source of confusion. But what exactly is it that the child has to learn? Wallon's description suggests that the child is able to recognize the image in the mirror as its father. Confusing is the disconnection between what it sees and what it hears. The child gets confused when the voice of his or her father does not come from the place where he or she is seeing him. This suggests that what the child in front of the mirror has to learn is not how to connect what it sees to other sensory input but rather to conceptualize the particular disconnection typical of the mirror. It is only after the child has learned to conceptualize the disconnection of what it sees from what it hears that the mirror image can appear as a convincing representation of his father, that it can appear to be 'seeing truth itself' in the mirror.

The members of the audience of *Double Track* are not children, but grown-ups who presumably already know the difference between mirror images and physical presences. Nevertheless, the architecture of *Double Track* somehow appears to evoke a confusion similar to the one described by Wallon. Similar, but not the same. In the case of *Double Track*, this 'mistake' happens even though everybody 'knows better'. The audience knows that they are watching a mirror image and that therefore, logically, the sound of the feet should come from the opposite direction. I already knew what the child still has to learn, but *Double Track* also made me aware of how I automatically responded in a way that aimed at preserving the illusion seen in the mirror at the cost of signals that suggested otherwise. Although the sound of voice and feet reconfirmed repeatedly the well known fact that I was watching a mirror image, it was very hard to grasp it as such.

The sound of the feet and voice of the dancer in *Double Track* causes a confusing oscillation between bodily disappearance and bodily dys-appearance. My response to it made me aware of my body as what Merleau-Ponty (discussed in previous chapter) terms 'a place of intertwining' that produces the unity of the world as well as a sense of self as seer in relation to it, even if this awareness manifests itself in the suggestion of being not involved. In *Double Track* what Leder terms the 'phenomenological condition of bodily absence' is produced and re-produced at the cost of denial of the 'sound track'. *Double Track* thus points to the necessity of conceiving of what we think we see as the product of an active and creative involvement of the body with its surrounding. The difference between my response and that of the child described by Wallon also points to something else, namely how cultural conditioning is part of how this connection is being made. Whereas the child looks in the direction of where it heard the sound of the father coming from, I tended to ignore the sound and placed a premium on the visual instead.

Mimicry, again

How to account for this active and creative involvement of the body in how we see what we see? How to account for this involvement in a way that allows for an understanding of the disembodied eye/I of modernity as the mode of looking of a body conditioned to see according to culturally specific parameters? How to do so in a way that can account for my response to the mirror/stage in *Double Track*, as well as the differences between my response and that of the child observed by Wallon?

I propose to do so starting from a concept that itself is emphatically about bodily response to what is seen, namely mimicry. Both Lacan, in his mirror stage essay, and Silverman in her reading of it, mention mimicry to account for ways in which the body can actively adapt its posture and movements to match an image of 'itself' seen from a point of view outside itself. Mimicry is a means of assuming an image through a process of bodily matching.[6] Silverman elaborates on this possibility in Lacan in order to describe how human subjects do not always wait passively and unconsciously for the gaze to 'photograph' them in the shape of a pre-existing image. On the contrary, they may give themselves 'to be apprehended by the gaze' in a certain way, by assuming the shape of either a desired representation or one that has come through less happy circumstances to mark the physical body' (Silverman, 1996, p. 201).[7] Yet, neither Lacan nor Silverman considers the possibility of mimicry as an active force at work in the constitution of what we think we see. This possibility is central to John Martin's elaborations on what he terms 'inner mimicry' and the experience of seeing modern dance.

According to Martin, movement is central to our first reaction to every object and circumstance we encounter. 'The only function of sense impression' he argues 'is to prepare the body for appropriate movement with relation to the objects reported upon' (Martin, 1939, 42). This does not mean that the body will make all the movements for which it is prepared. Based on past experience, we know that many noises that strike the ear are not indicative of dangers that must be fled from and as a result 'the actual carrying out of the movement for which the muscles are prepared is inhibited by this knowledge' (Martin, 1939, p. 43). The impulse is registered but short-circuited on the basis of past experience.[8]

Movement is central to our way of responding to what we are confronted with. In the first place because of this close relationship between sense impression and movement response as just described, in which 'movement' refers to the (possible) actions undertaken by the body in reaction to the situation it is confronted with. But this is not the only way in which movement is central to perception. As Martin observes, apart from being a response to the outside world, every movement-response to this outside world provides a wealth of information about our *own* bodies as well: sense organs in the tissue of muscles and in the joints register every change of posture and inform us about the where and how of the various parts of our body. As a result, every movement of our body contributes to a sense of self as also well, even if this information is not always conscious.[9]

Martin terms this 'movement sense' but what he describes overlaps at least partly with what others have termed proprioception.[10] With his elaborations on movement sense, Martin aims at a theoretical explanation of how what might be called our proprioceptive systems are part of our encounter with what we see. Take the following example:

> When we consider the weight of a log that we chance to see lying across the path, there is awakened in us a pattern of movement responses based on memory of previous experiences with the weight of objects, which prepares us through our movement sense for the muscular forces that will be brought into play and the energy involved in lifting this particular log. We need not actually lift it, therefore, to know that it is heavy, and approximately how heavy. The report made by the eye is sufficient to open one of many beaten tracks in our neuromuscular experience and associate this object with previous objects with which we have had contact. *When we pronounce the log heavy, then, we are actually describing not so much the log itself which we have not even touched, as the motor reactions which occur in our bodies at the sight of it.* (Martin, 1939, p. 45; my italics)

Martin describes how what is seen is experienced through one's own bodily parameters. Our bodily responses then are projected onto the log that as a result appears to us as 'heavy'. This intertwining of proprioceptive information and exterioceptive awareness is even more poignant in the perceptual mechanism Martin describes as 'inner mimicry', also discussed in Chapter 7. Inner mimicry does not refer to conscious or unconscious attempts to relate to the outside world but is a consequence of the way proprioception and exterioception are necessarily intertwined in which what we see and what we hear is always translated in our own present and active experience.

> If we look at a building with columns supporting a mass above, we shall form a definitive opinion as to whether the proportion is good or bad according to whether the mass that is supported seems easily supportable by the columns in question or too heavy for them. This reaction has nothing to do with any knowledge of architecture: it is a motor response. The building becomes for a moment a kind of replica of ourselves and we feel any undue strains as if they were in our own bodies. (Martin, 1939, p. 48)

Martin describes how the looking at the building involves a momentary mixing up of self and building. We become building, or the building becomes us, and this mixing up affects our conscious perception of the building. This process involves an active 'mimicking' of what is seen in which the motor responses of the body seeing connect this experience to previous experiences and thus awaken earlier sense perceptions and the feelings, emotions, expectations etc. related to them. All of this will become integral part of how this body seeing sees the building and, vice versa, how this body seeing sees for example this building will depend on the previous experiences of this particular body and the 'storehouse of associations' (Martin).

Perceptual systems

Unlike Silverman, Martin does not presuppose an initial separation between vision and proprioception. His account is more in line with a Gibsonian approach in which the senses do not function separately, but rather as *perceptual systems* (Gibson, 1966). Understood as active systems, the senses are neither passive senses nor channels of sensory quality, but ways of paying attention. Furthermore, understanding them as perceptual systems helps to account for the ways in which the sensors traditionally associated with different senses of sight, hearing, smell, touch, and proprioception interact and overlap in the constitution of our perceptions of the world. Understood as perceptual systems, these senses are not mutually exclusive or separate systems. Instead, they interact in the constitution of a world that is visible, audible, and touchable at the same time.[11]

Gibson's ideas originate in biology, but the philosophical implications of his ideas seem to be useful for visual theory as it has developed within the humanities. The senses understood as perceptual systems, allows for an understanding of subjectivity as the effect of an active engagement of various perceptual systems (which do so simultaneously) with what the body finds itself confronted with. Subjectivity is the effect of how these sense systems probe, map and bring together the results of their engagement with the world. The cultural environment is emphatically part of this. At this point, Gibson's notion of the senses as perceptual systems is in line with Crary's idea that not only seer and seen, but also what it means to see, have to be understood from the ways in which perceptual practices take place within, and are organized by, specific cultural and historical situations. Here (see note 10) the distinction between systems, information and awareness is helpful. Gibson's proposal to consider the senses as perceptual systems, suggests the possibility of understanding seeing as a process of bodily response and investigation, measuring, exploring through sight and hearing, as well as through proprioception and kinesthetics. The response of the seer is the product of a body as the place where these various perceptual systems intertwine; they probe the world around us. In this response, various information is produced, connected and mapped, matched and measured against expectations, knowledge and desire. This in turn can result in particular awareness of self and of world. How this happens will depend both on the physical possibilities of the body to engage with the world, and on the way this body is marked by culture. The response of the seer will depend on how the body has learned to perceive itself and the world around it according to culturally specific parameters, how the body has learned to behave, how it is marked by experience and the address of others.

'It's there, it's me'

Conceiving of the senses as perceptual systems affords an understanding of how seeing takes place in *Double Track* as an effect of probing and combining through various sense systems simultaneously. What appears as visible world is the effect of how various informations produced by the perceptual systems are connected as a result of which the world as we perceive it is produced as both visible, audible, tangible etc. and all of this in relation to an 'I' as the origin of perception. In this process, various perceptual systems are involved, and it is through these that the body produces what we think we see in relation to a point of view, that is, in relation to a position to which the terms 'here', 'there' and 'my' are keyed.

Both this position and the world emerge as the result of our synaesthetic response to what we find ourselves confronted with. This response provides us with a perspective on the world in relation to a point of view. This point of view is not a fixed and static position as in a perspectival drawing, but constantly moving as we are moving along with what we find ourselves confronted with, repositioning ourselves in relation to what we find ourselves confronted with. This point of view or position, therefore, is not something given or imposed, but results from an interaction. It is our response to the address we find ourselves confronted with that provides us with a perspective on what we see. This brings me back to Damisch's observations on the analogy of deixis and perspective quoted at the beginning of Chapter 2:

> Perspective is not a code, but has this in common with language that in and by itself it institutes and constitutes itself under the auspices of a point, a factor analogous to the 'subject' or 'person' in language, always posited in relation to a 'here' or 'there,' accruing all the possibilities for movement from one position to another that this entails. (Damisch, 1995, p. 53)

It is not the perspectival construction in paintings or drawings that positions us but rather, such constructions present an image of how, as Damisch puts it, 'the subject is caught, manoeuvred, captured inside the field of vision, and how painting can deliberately exploit it to captivate the "subject" in a relation of desire, but one that remains enigmatic' (Damisch, 1995, p. 46). The place or position of this subject, therefore, is not the geometrical point defined by optic geometry. Rather, the subject moves about within the field of vision, positioning itself in response to the address presented. In this process, the 'I' as deictic marker of this 'place' marks the point of view emerging from our perceptual response to the address we find ourselves confronted with. It marks the position of origin of the perceptual perspectives opening up from the encounter of my body with the world. 'I' is the position in relation to which the terms 'here', 'there' and 'my' are keyed while at the same time, as Benveniste (1971) points out, this 'I' is an 'empty' sign. It is a position in discourse. Its meaning lies in how it is used to set up a relationship between 'I' and what is expressed in discourse or between 'I' and a you etc. Furthermore, the 'I' functions as a shifter between various instances of discourse, connecting successive moments and this way producing subjectivity as something that surpasses these individual instances while at the same time allowing for a radical discontinuous notion of what subjectivity is. Subjectivity consists in the succession of these individual moments of positioning. The analogy as pointed out by Damisch, of deixis in language and perspective, suggests the possibility of a further expansion of this discontinuous notion of subjectivity to include positions as subjects of vision as well. Positions that, as my analyses so far have demonstrated, always involve more than what is there to be seen. Positions that are not unitary, but always already conglomerates of positions emerging from the confrontation with an address that involves various perceptual systems simultaneously and in which perceptual positions resonate with discursive positions up to the point they become indistinguishable. The self-reflexive performative gestures of performances like *Artifact, Holoman; Digital Cadaver* or *Double Track* illuminate how such perspectives interfere with one another, and how they are the product of senses that are cultured to perceive in some ways rather than others. They suggest the possibility of a further radicalization of Benveniste's understanding of subjectivity as discontinuous and entirely relational, moving towards and understanding of bodies capable of complex experiences that result from interferences, resonances and even contradictions between the various positions emerging from the interaction between seer and seen.

Furthermore, the analogy pointed out by Damisch suggests that like perspective in painting, the perceptual perspectives implied within the multimedia address that is the theatre may present us with an image of how the subject is caught, manoeuvred, captured in the field of vision. How the coherence of this world as perceived and of the self as experienced in relation to this world, is not something preceding perception but the product of it. In this context, the Lacanian mirror stage appears as an account of a culturally and historically specific mode of processing experiences, producing a particular type of self versus world and vice versa. A type of processing, furthermore, that privileges the visual as point of reference in relation to which the world is constituted as a perceptual unity that is both visible, and tangible and audible etc.

10
Welcome to What You Think You See

> Good evening. Remember me? Good. Now try not to forget what you are seeing, and you will think what I hear. Try not to remember what I am doing and I will say what you thought. Try not to forget what you are hearing, and you will see what I think. Try not to remember what I am saying, and you will hear what you do. Try not to forget what you are doing and you will hear what you say. Try not to remember what I am saying and I will see what you think. Do you see what I mean? (from *Artifact* by William Forsythe, 1984)[1]

Remember her? The woman in the historical costume from *Artifact*? She who invited the audience to leave its seats and 'step inside' the performance, to live it from her perspective? This woman whose appearance in my text marked my invitation to you to 'step inside' my theoretical engagement – that is, my engagement with the enigmas that characterize our perception of the visible world. This world's appearance that (as I have argued) is the outcome of a process of seeing that always involves reading, even if it seems to be 'just looking'. A world where the seer is involved in a process that cannot be understood solely either from what is implied by what is seen, nor from the constitution of the seer. The seer is always involved in what is seen, just as what is seen always implicates the one looking. This seer is a body involved, simultaneously, through all the various senses with what is seen, a body responding to the situation made explicit by the woman in *Artifact* as she bids the audience a cordial 'Welcome to what you think you see'.

In her short monologue, the woman theatricalizes the relationship between her statement and the (presumed) response of an addressee, highlighting how 'I' and 'you' emerge as positions within processes of cognitive perception, or perceptual cognition, as these processes occur between the seer and the seen. As the viewer's comprehension of what the woman means by 'what you think you see' builds, 'I' and 'you' emerge as mutable agents, connecting various perceptual and cognitive acts. Understanding what she means is thus presented as the outcome of complex synaesthetic processes, inseparable from thinking, remembering and forgetting. And, the capacity to identify, as an analytical strategy, the various positions of 'I' and 'you' that arise from the interaction between address and response emerges as helpful in understanding the complexity of what is involved in structuring what we think we see.

In this chapter, I follow the woman's suggestion. I propose a method for analysing how visuality 'takes place' in the theatre, starting from ways in which the theatre addresses the seer, and how the cognitive-perceptual perspectives thus presented suggest specific relationships with and stances towards what is seen. And also how, as a result of these specific relationships, the seer, despite the fact that (to follow Martin) he or she may seem, to all outward appearances, to remain quietly seated, is nevertheless invited to move along, thereby producing a choreographical inscription (Lehmann) in choral space. Following this, how 'moving along' produces what the seer thinks he or she sees primarily in relationship to the stance of an 'I' seeing; and how the various stances emerging from this interaction may resonate, interfere with, or contradict each other. Further, how stories, associations and interpretations may be invited, within which self and other conflate. I will also demonstrate how various concepts previously introduced also support such an analysis of visuality as it occurs in the theatre. To do so I will analyse scenes from four different versions of the same opera: Wolfgang Amadeus Mozart's 1782 *Die Entführung aus dem Serail* (*The Abduction from the Seraglio*).

The theatrical practice of staging the same play or libretto over and over again has a lot to offer when it comes to understanding the relationship between what we see and the cognitive perceptual practices invested in seeing. Staging something is providing a perspective on what is shown. It means showing something in relation to a point of view, both in terms of spatial relations and in terms of the symbolic spaces opened by discourse. In the case of a staging of a play, not only the words of the text guide how we see what we see. Every visualization also implies an interpretation of the textually represented world. Staging means determining what things look like and deciding what seems convincing (or not), who is to be sympathized with and for what reason, what is acceptable – or not – to embody or show. Staging, therefore, is a gesture of exposure in the sense described by Bal (1996a), that is, an exposure that produces the events, situations and characters described in the play or libretto in relation to a particular point of view. The success or failure of such acts of exposure is not a measure of what one person making the gesture wants to say, but what a community and its subjects think, feel or experience to be the consequence of this exposition; how what is shown is recognized (or not); and, how the address presented to seers meets their expectations, assumptions, desires and presuppositions. For this reason, the theatrical practice of staging and restaging the same texts over and over again presents an interesting arena within which to analyse cultural and historical specificity of the relationship between visions presented and point of view implied by these visions, and contemporary indices of transformation as well as constants in culturally specific modes of looking.

'Turkish' delights

Die Entführung aus dem Serail tells the story of a young Spanish woman (Konstanze) who, with her English maid (Blonde), and servant (Pedrillo), is abducted by pirates and sold to a Turkish man (Bassa Selim). He adores her but she has already promised her heart to Belmonte, her fiancé, and is determined to save herself for him. She realizes, however, that she will not be able to resist Bassa Selim for long. Asking for his compassion, she begs him to give her one more day to mourn the loss of her lover. After that, she promises, she will give herself to Bassa Selim. Bassa Selim grants her request. In the meantime, with the help of Pedrillo, Belmonte manages to enter Selim's house and designs a plan to take Pedrillo and the two women back home. Pedrillo pretends to want to make up with Selim's servant Osmin and suggests a drink together, feeding Osmin wine containing a sleeping potion. However, when the four of them try to leave the house in the middle of the night, Osmin is less easy to fool than they thought, and he captures them. Osmin, who had distrusted them from the start, now finds his distrust justified. Their fate appears to be sealed, even more so when Belmonte turns out to be the son of Selim's arch enemy. Belmonte and Konstanze, certain death is near, once more declare their love and bid each other farewell. But then the unexpected happens. Bassa Selim returns deceit with compassion and sets them free. All celebrate the glory of Bassa Selim.

Die *Entfuhrung aus dem Serail* is one of Mozart's early operas. It belongs to the Singspiel genre, which means that some texts are spoken. In *Die*

Entführung aus dem Serail one of the characters, Bassa Selim, does not sing at all. The story develops through the opposition of Old Europe to its Turkish other. The title *Die Entführung aus dem Serail* evokes images of adventure and abduction and the promise of a happy ending. The references to Turks and Turkish played upto the exotic fantasies about cultural otherness of Mozart's contemporaries, an otherness that was perceived as both threatening (a century earlier the Turks had besieged Vienna) and exciting (the harem as projection screen for sexual fantasies). Such Turkish motifs were very fashionable when Mozart wrote this opera, but later became the basis for much criticism. Many have pointed out how 'Turk' and 'Turkish', both in this opera and in other musical dramatic presentations of the period, serve as a label for all kinds of fantasies set in the exotic East or involving characters from the East, often reducing them to caricatures (Osmin) or unrealistic idealizations (Bassa Selim).[2] Such works are examples of what Edward Said (1985) has famously termed orientalism: a system of representations forced upon the East, inscribing it within Western ideological constructs.

The implication of Said's argument is that if we want to understand and criticize the meaning and implications of orientalist representations of otherness, it does not suffice to point out that they are wrong in the sense that they do not show the other as he or she really is – which of course they don't. Rather, if we want to understand the implications of such representations, we have to take into account the desires, presuppositions and anxieties invested in them. We have to consider how such representations respond to a particular point of view. Whose point of view is this, and how does representation mediate when this point of view is the one taken? In the following, I will start by taking scenes from four different versions of *Die Entführung aus dem Serail*. I read these scenes as four different visualizations of what is at stake in this opera, each providing the audience with different points of view. These are partly given within the construction of the libretto, a construction that, like the dramatic frame as described by Lehmann, acts as an invisible logic directing our attention to what is shown on stage. Yet, as I have argued (in Chapter 4), when staging a text or libretto, a director must literally produce a vision of the text, deciding what the audience will see. This involves deciding on the relationships between words and images, the verbal and the visual and, in the case of opera, the music. Staging, therefore, means producing a reading of the libretto and this reading, in its turn, addresses the audience in ways that suggest points of view from which to see what is presented to them. My concern is how this happens in these four versions and to what end.

The four stagings are, first of all, Jonathan Miller/Opernhaus Zürich (2003)[3] and Massimo Teoldi/Teatro della Pergola (Florence, 2002),[4] both faithful to the original libretto and musical score (and both available on DVD). I will compare scenes from these with Ramsay Nasr/Muziektheater Transparant (Antwerp, 2006)[5] and Ibrahim Quraishi/ Schauspielhaus (Vienna, 2006)[6], both of which took more liberties with libretto and score and presented adaptations that implied a critical stance towards the (presumed) orientalist character of the original. In my analysis, I will focus on several questions, as follows: How do these performances invite us to 'step inside'? How do they visualize otherness? How do the productions mediate in establishing a relationship between the viewers and the presented visions of otherness? What is the effect of this mediation in our understanding of the otherness presented by each performance? How do these performances invite us to understand the relationship between what we see and what we hear? Where does this put us as viewers in relation to what we see?

Meet the other

Die Entführung aus dem Serail begins with the arrival of Belmonte at the palace (Serail) of Bassa Selim, where he is confronted with Osmin, Selim's servant and guard. Belmonte tries to talk to Osmin because he wants to know if (1) this is Bassa Selim's palace and (2) if his former servant Pedrillo is working here. Osmin is unwilling to help and makes clear that their (meaning Pedrillo and Belmonte, the Europeans) kind is not to be trusted, that they will not fool him, and that if it was up to him (Osmin), he would hang them right away and put their heads on stakes. The scene thus puts Osmin in the position of the one who is prejudiced against Europeans and Belmonte in the position of the object of his prejudices. The scene gives a demonstration of the mechanism of 'othering' through the assumptions that guide the culturally determined view via a situation in which the tables are turned. The other is made to look back at what, the original audience of this opera, most likely has appeared as self. At the same time, the grotesqueness of Osmin's behaviour neutralizes the threat this scene might present to our understanding of self, suggesting he is not to be taken seriously. There seems to be no reason for his unfriendly behaviour, which seems rather exaggerated and unnecessary. We are invited to conclude that this man is unreasonably biased against Europeans.

This is the situation of the first scene, in brief, as written in the libretto. What does it look like on stage? In both Jonathan Miller's staging with the Opernhaus Zürich and Massimo Teoldi's staging with the Teatro della Pergola, Osmin's appearance confirms the orientalism for which this opera has been criticized. In both versions, his 'Turkishness' looks like some undefined fairytale Eastern-ness, with Osmin wearing a colourful costume of harem pants or a skirt, a scarf around his waist, a turban on his head, pointed shoes or boots, a dagger or sabre, and heavy brows and a dark beard. His looks confirm what we 'know' about the Orient from popular representations of Scheherazade and the 1001 Nights and numerous other paintings and drawings as they are part of the cultural imaginary or visual unconsciousness (the screen), thus confirming the gaze that directs the Western look on its Eastern other.

There are, however, also important differences between these two versions, differences that are telling when it comes to how the performances invite us to look at, and relate to, these orientalist appearances. In the Zürich version, the curtain remains closed during the overture. Belmonte enters the stage just as the curtain opens for the beginning of the first act and he starts his first aria, singing 'So this is where I am to find you, Konstanze'. As was the case with the woman in *Artifact* (Chapter 2), Belmonte's appearance marks the beginning of the performance. He is literally our (the audience's) point of entrance into the imaginary world, taking us with him on his quest for Konstanze. He finds himself in an enclosed space consisting of three blue and white walls, a typical box-set stage. The stage design is modest, lacking detail, bringing to mind both Moorish-like architecture and Western architecture of Mozart's time. One palm tree in the middle is the only thing explicitly exotic.

'But how will I get into the palace? See her? Speak to her?' Belmonte wonders, directly addressing the audience, while Osmin enters behind him through a door in the back. The staging turns Belmonte's text into a moment of external theatrical communication that highlights the relationship between stage and auditorium. Like the woman in *Artifact*, Belmonte acts as an internal focalizor. As in *Artifact*, the explicit address presented to the audience prevents automatic absorption, and instead highlights the theatricality of this situation wherein we, the audience, are invited to imagine another world behind the proscenium arch.

Osmin appears unwilling to answer Belmonte's questions. At first, he does not give a direct reply, 'answering' with a song: 'Once you have found a sweetheart who is faithful and true, then smother her with kisses, make her every wish come true, be her comfort, be her friend.' When Osmin stops singing momentarily, Belmonte turns to the audience again and pointing to Osmin behind him, says: 'Perhaps this fellow will tell me what I need to know', thus again explicitly implicating the audience in the events on stage. Osmin continues his song with a warning: 'But to keep her nice and faithful, put her under lock and key', for, as he explains, women are very easily attracted by other men. Although Osmin does not explicitly say so, his behaviour suggests he suspects who Belmonte is and what he is coming for, and it looks like he might be teasing him with this song, delaying his quest to save Konstanze from other men, from the Other. Only when Belmonte gets very irritated does Osmin answers his questions, yet still in a way that makes clear that, according to Osmin, Belmonte is not in the position to behave the way he does.

This staging of the scene shows the interaction between Belmonte and Osmin as one of mutual irritation. Osmin may be prejudiced against the Western others (Pedrillo most of all) but Belmonte's demanding behaviour towards Osmin is not very respectful either, especially given the fact he is the stranger in Osmin's place. His way of talking to Osmin, a little too slowly and excessively articulated, suggests that he thinks Osmin is a bit stupid. Belmonte elevates himself to a position of superiority, inviting us, the audience, to share this position. The scene develops into a power play between the two with the audience as third party to the contest. Both characters repeatedly and explicitly address the audience, as if it is into the eyes of the audience that they are looking for confirmation of their side of the story.

Although Miller's staging of the scene sets up a connection between the audience and Belmonte from the start, the explicit theatricality prevents automatic alignment of one's vision with the character. Instead, the mechanisms by which this address proposes a particular relationship to what is seen are illuminated. And by extension, the theatrical game between the characters reads as illustration of how staging responds to and plays with expectations, desires, and presuppositions of viewers. From this perspective, the audience is invited to notice the artificiality of Osmin's looks, the fact that he is dressed up as for a costume party, his dark skin and beard clearly made-up. Belmonte, in contrast, looks like a perfect copy of portraits of the young Mozart. This dual presentation might be read as indication of what point of view we are invited to adopt and how the vision of the other presented to us is mediated by this particular cultural-historical perspective. Here, the imaginary, onstage world is presented as the product of historical imagination, and as explicitly theatrical: fiction. What we see is the theatrical imagination of another time. This reading is supported by the visual relationship between the stage design and Belmonte's appearance.

Although the 'palace' on stage is supposed to be Osmin's world, where Belmonte is the 'other', it is Belmonte whose looks match the stage design. As a result he looks more 'natural' (a 'naturalness' that is also confirmed by his behaviour), whereas Osmin looks strangely costumed.

All of this is rather different from Teoldi's version with the Teatreo della Pergola. Here too, Belmonte is the internal focalizor who invites the audience to step into his shoes and enter the imaginary world with him. As with Miller's version just described, the scene is constructed as to invite the audience to enter the imaginary world with Belmonte. Here however, it is not Belmonte's stepping from the wings that sets the imaginary world in motion. In this production his arrival is visualized as a long journey within the imaginary world behind the proscenium arch. He already makes his appearance during the overture, presenting us with a vision of what happened before the actual opera begins. Far in the distance, a small ship crosses the stage from left to right and right to left etc., each time more downstage, and each time a little bigger. Finally a boat arrives at the very front, with Belmonte in it. The scenery changes to the front of a palace and at this point Belmonte sings his aria about hoping to see Konstanze again here.

Belmonte travels towards us from the endless depths of this other world, to meet us downstage, in front of Bassa Selim's palace, and to take us inside with him. This other world is staged in such a way as to invite us to look at it as having an autonomous existence, behind the proscenium arch, a world that extends beyond the limits of what we can see. In this world, Osmin too appears not just through a stage door, but from somewhere far behind. We hear him singing long before we see him entering. Again, the staging invites us to imagine the stage as a world that extends beyond the set as we, together with Belmonte (literally looking over his shoulder), listen to Osmin's voice coming closer. Sound is instrumental in setting up this relationship between seer and seen, supporting the illusion of a world extending beyond what we can see and providing us with a position as auditory subject in which the 'I' of 'I hear' conflates with the position visually marked by Belmonte.

It is Osmin whose costume and looks blend in, the stage design embracing him as an integral part of a strange other world that we (with Belmonte) enter. This not only affects the power balance between the two characters (granting Osmin the advantage) but also 'naturalizes' Osmin's otherness. Instead of highlighting the relationship between the world as presented by the opera and a particular seer, this staging presents the opera as a window to another world, one existing independently. This independence turns the gesture of exposure performed here into a constative one: this is how they are. Within this theatrical logic, Osmin is what he looks like, a brutal and violent man who, totally absorbed in his own world, does not initially even notice Belmonte. As described above, he starts singing his song long before he even enters the stage (and long before he could possibly have noticed Belmonte) and continues (once onstage), while feeding a crocodile. Belmonte does notice Osmin and tries to attract his attention, while emphatically denying the presence of the audience, thus setting the code that we, the audience, are not there. The behaviour onstage suggests the performers are not aware of being seen. Within this logic, we, the audience, are invited to forget our position as seer in relation to this world and, by extension that the construction of this world responds to the presuppositions, expectations, and desires of a culturally and historically specific seer.

Total abduction

These culturally and historically specific expectations, presuppositions, and desires are how Osmin and his world can become explicit commentary in Ramsey Nasr's version of *Die Entführung* for Muziektheater Transparant. In this staging Belmonte enters the stage somewhere during the overture, and finds himself confronted with a dark red rectangular object. He looks at it and wonders where the hell the f***** palace is, the palace where he is supposed to meet Konstanze. Expecting to see a palace, he is confused when no palace appears. Later we will learn that the dark red object is Bassa Selim's palace, although it certainly does not look the way Belmonte (or we in the audience) had expected.

Nasr's staging presents the audience with an internal focalizor, who knows the opera and therefore expects certain things – which don't materialize. The stage world confirming clichés at work in our vision of our contemporary Islamic other, does not fulfil his (and probably our) expectations of some imaginary Orient. The first characters to appear on stage are four women dressed in wide greyish dresses and veils, leaving only their eyes uncovered. The looks of the Europeans are equally clichéd. Belmonte, in white captain's uniform, could have walked straight out of *The Love Boat*, or an operetta (Figure 13). Blonde's white servant's dress is reminiscent of a nurse from a hospital romance. Furthermore, all the Western characters are presented as internally split. They are performed by two people: one actor and one singer. The singers are the younger versions of the characters and perform Mozart's idealized vision of love and romance, whereas the actors, representing

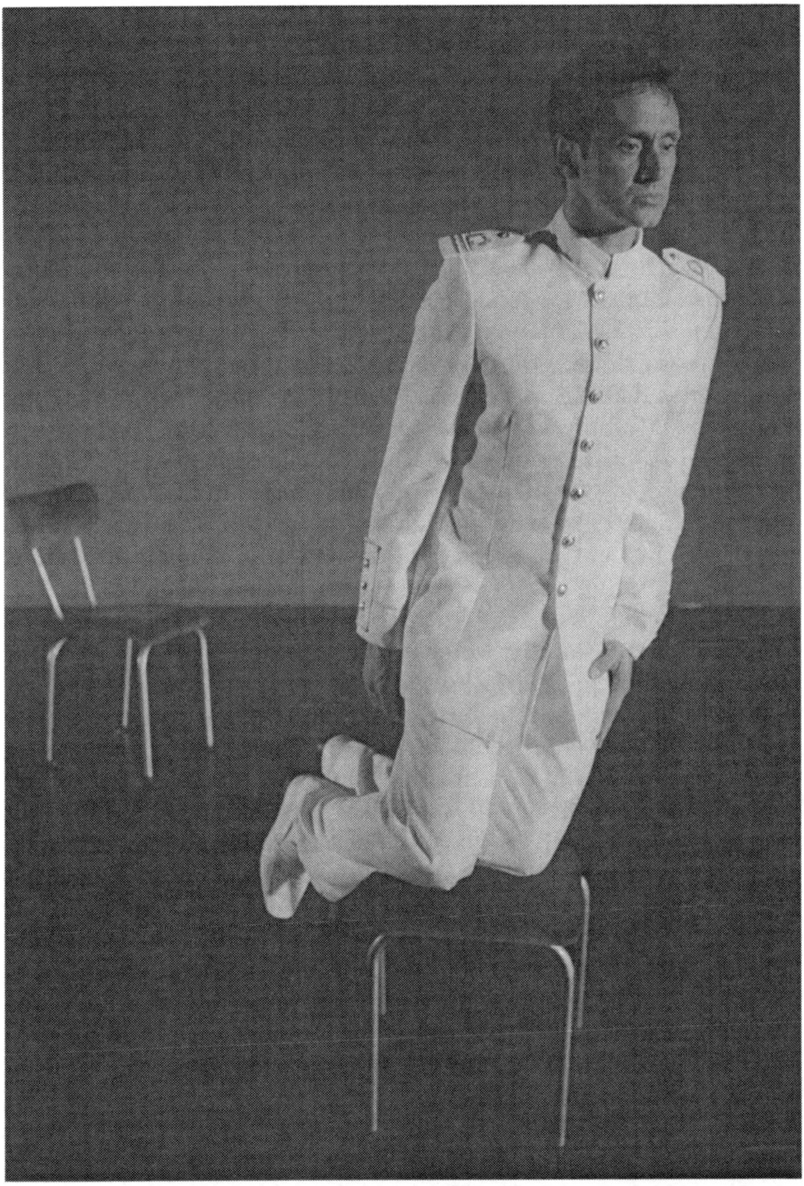

Figure 13 Photograph by Herman Sorgeloos showing Christian Baumgärtel as Belmonte (the singer) in Ramsey Nasr's *Een Totale Entführung*. Reproduced with permission of the photographer.

older versions of the same characters, get involved in fights and disagreements, misbehave or do not behave according to the plot, even step out of their roles. The staging suggests that the opera singers are the more ideal mirror images, providing the older actor-versions of the characters with an entrance into an imaginary other world. This turns Nasr's production into a meta-theatrical commentary on the visions of self and other provided by the opera, and on how these visions play upto the desires of viewers.

This suggestion is made explicit at the beginning, when Belmonte/the actor finds himself lost on stage in a world that does not confirm his expectations, looks around disturbed, and finds what he is looking for in Belmonte/the singer. While seated in the auditorium, the younger man begins, thus making his first lines ('So this is where I am to find you, Konstanze. My only joy. Heaven hear my plea. Restore my peace of mind)' into an expression of what the audience might be looking for, the kind of desire to which the world represented in the opera (including the characters) responds, or should respond to. The singer eventually leaves his seat in the auditorium, literally climbing onstage to become part of the imaginary world there; audience member becomes Belmonte. Belmonte/the actor helps Belmonte/the singer to take off his jeans and get into another white captain's costume. Through all of this, he continues singing the opening aria.

Who is this seer wanting to step inside and become part of the opera world of *Die Entführung aus dem Serail?* A scene halfway through the opera contains a very explicit commentary on this question. Pedrillo fools Osmin into believing he wants to be friends, and seduces him into drinking wine to which Pedrillo has added a sleeping potion. Within the libretto, this scene is a kind of *deus ex machina*. It is unconvincing that Osmin would suddenly believe in Pedrillo's friendliness, and it is also odd that Osmin would suddenly want to drink alcohol. Drinking turns him into a fool, which he is not – he is the one who rightfully distrusted Pedrillo and Belmonte from the beginning, suspecting what they were planning. Furthermore, it remains unclear how, after this narrative twist, it is possible that the drunk Osmin is able to catch the lovers when they are trying to make their escape.

Nasr makes the scene during which Osmin is fooled by Pedrillo into a 'play-within-a-play'. The inner room of Bassa Selim's palace becomes the stage on which Pedrillo and Osmin perform in extremely exaggerated vaudeville style, wearing costumes that are grotesque versions of the orientalist clichés they themselves represent (Figure 14). First, Pedrillo helps Osmin into a costume, also giving him a piece of paper with his lines. Then he speaks to the audience, saying 'Come and see: the story of the smart Christian, or the downfall of the miserable Muslim. Tonight in a theatre in your neighbourhood.' What we see are two characters, part of the fictional, onstage world, who now perform a highly theatricalized version of a scene from the play of which they themselves are part. The play-within-a-play is explicitly staged by a Western hand, and equally explicitly casts the Eastern character as the dolt outwitted by Western cleverness. The multiplication of the theatrical frame undermines any stable point of view in relation to what is seen. It also directly confronts the seer with the desires and presuppositions to which this scene (and by extension, the opera) both responds to, and plays upto. The explicit theatricality and extreme exaggeration allow for a certain distance from what is presented, and invites a reconsideration of what is at stake in such representations. The framing provided by Pedrillo's text not only redirects attention towards what is at stake in this particular scene, but invites a similar mode of looking with respect to the other clichés and prejudices referred to and played with, time and again, throughout the performance.

Nasr's adaptation was titled *Een Totale Entführung*, which translates as 'a complete abduction'. The title brings to mind Adolf Hitler's question

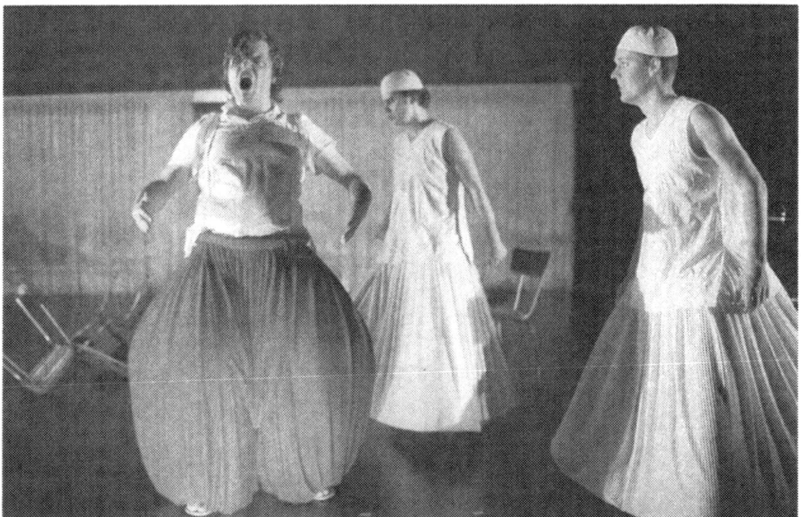

Figure 14 Photograph by Herman Sorgeloos showing Stefaan Degand, Tom de Wispelaere and Peter Gijsbertsen in Ramsey Nasr's *Een Totale Entführung*. Reproduced with permission of the photographer.

'Wollt Ihr den Totalen Krieg?' the question with which he challenged his audiences, encouraging their desire to have him lead them into the war of wars, total war, which is what Nasr gives us. He shows both sides prejudiced against the other, their visions clouded with all kinds of assumptions. Nasr opposes the orientalism that characterizes Western visions of the East with equally vicious occidentalism at work in how Easterners look at the West. Yet as a visual text his production demonstrates that the situation is not symmetrical in terms of how this performance relates to a seer. For, although the behaviour of the European characters towards the 'Turkish' ones indeed at some points can be called orientalist and the behaviour of the 'Turkish' characters (especially Osmin) occidentalist, the orientalist characteristics of Osmin and Bassa Selim are not projections of the European characters, that is, they are not the result of how the European characters look at them. Osmin and Bassa Selim look the way they do, not because of the orientalist assumptions of other characters on stage, but because of the way they themselves respond to the assumptions, presuppositions and desires of a seer outside the representation. This is what the play-within-a-play strategy is used to comment on.

The play-within-a-play questions this external focalization as it is implied within the construction of the play. By extension, it also questions what is up for grabs in what, from this point of view, appears as a convincing representation of otherness. The staging invites reconsideration, pointing to the implicit assumptions and presuppositions at work in what otherwise might seem self-evident or unproblematic. Time and again, explicit theatricality is used to expose elements of this relationship: between what is seen and the subjective point of view. Time and again, except for the final scene.

After more than two hours of exaggerations, role-playing, screaming and fighting, the actors take off their costumes. The singers (the idealized versions of the characters) sing their final songs and leave. Bassa Selim announces that he has decided to let the lovers go, and the actors leave the stage also. Only Konstanze lingers. She seems to find it difficult to leave Bassa Selim but, in a short monologue, argues that it is better this way. He here, she there, each in their own place. Bassa Selim remains with Osmin, who returns to tidy the stage, replacing chairs – things he was doing when the performance began. In line with the deconstruction of the theatrical fiction presented so far, this might be read as sign of the fact that they are only characters, inventions that have no existence outside the theatre. They are constructs of European imagination, invented to play out fantasies of superiority and exotic adventures on a European stage. They cannot leave the theatre.

Yet, the scene that ends the opera suggests something else. Bassa Selim, alone now and wandering around on an empty stage, singing an Arabic-sounding song, suddenly appears as more real and authentic than everything we have seen previously. All explicit theatricality removed, he is there as simply himself, and it is an absorbing spectacle. Instead of confronting us with the inevitable screen through which we perceive the other, the final scene suggests it might be possible after all to see (and to show) otherness as it is. This final scene thus presents a break with the preceding logic of the performance, a break that, like the scene termed 'the navel' in *Bas and Elze Dance* (Chapter 5) allows for two different, mutually exclusive readings. This scene can be read as analogous to what Barthes describes as the punctum, the 'lo', the 'there it is' moment. This is the moment that allows for direct contact with that which is actually there; a moment in which we finally perceive the other as he or she really is. Therefore, the amount to which Bassa Selim is convincing to us as viewers is not dependant on the way his appearance responds to our desires, presuppositions and expectations, but is instead located firmly over there, in his presence, in the fact that he is what he is. This reading, however, contradicts the claim made by the rest of the performance, namely that what we take for the other is in fact our own projection. The appeal presented by the final scene suggests that it would be possible after all to undo such cultural perspective at work in our gaze. This final scene presents a rhetoric of deconstruction to foster a sense of increased authenticity, of immediateness. The thing is brought back to itself in a manner analogous to the inseparability of dancer and dance in the rhetorical reading of Yeats's poem *Among School Children* (Chapter 5). While the performance up until this final scene time and again points to the discursivity of our ways of seeing, the removal of the explicit and exaggerated theatricality endemic to the rest of the opera suggests that the closeness to the stage world experienced in this final scene is not due to Bassa Selim's appearance as a representation capable of confirming deeply held beliefs and desires so as to make us (the audience) forget our position of seer in relation to what is seen (which is the relationship stressed throughout the rest of the performance), but is the outcome of finally seeing him as he is. What he is (or at least, what the performance suggests he is) is the glamorous, sensual yet inaccessible other, distancing himself from the squabbling of the other characters fighting their fights, elevated and alone in his tragic heroism. One might argue that Bassa Selim's appearance in this final scene confirms some of the very same orientalist assumptions that Mozart's original has been criticized for.

This final scene also presents an interesting example of the absorptive qualities of music and of the singing voice and how the positions emerging from the relationship with what is heard mediate in how we relate to what we think we see. It is with his singing to this Arabic-sounding music that Bassa Selim, in Mozart's version a speaking-only character, draws us close, as did the other, Mozart-singing characters during the rest of the opera. His singing, and the music, turns the space into his space. We seem to feel his feelings, moving along with him as he wanders the stage. He provides us with an experience similar to those that, within this staging, the other characters seem to be looking for in the imaginary world of the opera – experiences they long for, dream of and desire; dreams, desires and fantasies these characters allow us to imaginarily play out on stage. Yet, whereas the absorptive qualities of singing and the illusion engendered through the music of shared emotions are clearly and repeatedly framed throughout the opera as theatrical conventions playing into the desires of the audience (a position represented by the actor-characters on stage), in this final scene, such framing disappears. Selim appears as the answer to melancholic desires for a noble, pure and sensual other, which is precisely the kind of seduction that is also thematized in the plot of *Die Entführung aus dem Serail*, where these seductive capacities of Bassa Selim present a threat to Konstanze's faithfulness to Belmonte.

Hearing and feeling

Nasr's adaptation uses the maid Blonde to comment on these seductive capacities. Blonde suggests that, actually, Konstanze would be much better off with Master Selim than with Belmonte. Selim has this, 'you know, kind of mildly primitive, African, Mediterranean thing'. Her conclusion: 'It's a pity he's Muslim, though'. The original opera too allows for a reading in which Selim's oriental character appeals not only to the appetite of the audience, but is equally seductive to Konstanze. At this point, the libretto allows for different interpretations and these different interpretations imply different ways of relating the visual, the verbal and the auditory as presented in the libretto and score.

Crucial here is the scene in which Bassa Selim reminds Konstanze of the fact that this one day she asked for is almost over, and that she has promised that tomorrow she will be his. Konstanze answers that she will never love him, she would rather die. He answers that he will not let her die, but suggests he might torture her. She then responds with an aria, stating that she is not afraid of such threat and that in the end death will set her free. In Teoldi's version, Bassa Selim is quite awkward and his remark that he will torture her seems to have slipped out due to frustration over the fact she does not surrender to him. Konstanze gives no indication at all of being even slightly attracted to Selim, and we are invited to believe that indeed she will resist until she dies. The music accompanying her aria, alternating between a soft and fragile melody and a powerful, almost aggressive one, echoes her alternating feelings of sadness at her tragic fate and determination to never give in to Bassa Selim's power.

In Miller's version, on the other hand, Bassa Selim (given an excellent performance by Klaus Maria Brandauer) is not in the least aggressive or awkward but on the contrary, very charming and friendly. He seems to be playing with Konstanze's feelings rather than threatening her, and his performance suggests he means it when he says that he wants to win her heart and not force her. The remark about torture is made teasingly, saying what she expects him to say, or even what she wants him to say. For, in this version, the behaviour of Konstanze suggests she is very much attracted to Selim. What threatens her is not the prospect that he will force her to be his but that she is not far from giving in to someone she should not give in to.

This reading of the relationship, as given in the libretto, between Bassa Selim and Konstanze invites a different way of relating the music to the scene seen and the words sung. Whereas Teoldi's version suggests an understanding of the music in terms of, to speak with the woman from *Artifact*, the promise that 'you will hear what I (Konstanze) feel', or even, after Martin (Chapter 7) that you (the audience) can feel what I (the character) feel, Miller's version invites a reading in which the perspective provided by the music alternates between the positions or 'I's on stage represented by Selim and Konstanze. Here, the soft melody resonates with the soft seduction of Selim's address to Konstanze, with the more powerful music corresponding with Konstanze's insistence that she will resist Selim.

In both cases, music acts as a kind of focalizor, inviting the audience to take up a position in relation to the situation represented on stage. In both cases this is the position of a 'you' who can hear what an 'I' on stage feels and, through the music, is addressed with the invitation to 'take up' the position of this 'I'. Yet, what position this actually is cannot be understood solely from the music. It results from the constellation of positions emerging from the interaction between audience and performance. Furthermore, in both cases, the visual appears as the point of reference to which the auditory is keyed and in both cases, the music directs our attention so as to confirm us in what we think we see. At this point, Ibrahim Quraishi's version of *Die Entführung aus dem Serail* presents a very different example of how music can be used to manage attention in a way that prevents absorption, and to question the vision presented by the opera.

House of mirrors

Quraishi opted for the part of Mozart's title that was left out by Nasr (Serail) and titled his version *Saray. Mozart alla Turca*. Instead of criticizing Mozart for doing Turkish music his own way and transforming it into something that is only vaguely reminiscent of real Turkish music, Quraishi follows Mozart's example and takes the liberty to do Mozart's music in a Turkish manner. Composer Serdar Yalcin adapted Mozart's opera for a combination of piano, traditional Turkish instruments and electronics. Part of the result is recognizable as Mozart's composition, yet sounds very different from conventional orchestrations. This new sound confirms that Mozart's 'Turkish' music has little to do with Turkish music. Performed on Turkish instruments, Mozart's 'Turkish' music does not sound Turkish at all, but very much like Mozart, performed on unfamiliar instruments. Rather than absorbing the audience in music that sounds comfortably familiar, this version de-familiarizes the well-known sound, inviting the audience to listen in a different way, highlighting the structural characteristics of the music instead of directing attention to the execution of the well-known melodies and arias.

Similarly, Quraishi's staging highlighted the structural characteristics of the libretto, inviting reflection on this structure and its implications rather than inviting identification with the woes and worries of the individual characters. More than staging Mozart's opera, Quraishi put the opera on stage. His version highlights how the narrative construction of Mozart's opera evokes a reversal of positions, and how this reversal creates a tension between the supposedly self-evident story of the woman who has to be saved from the Turks, and a representation in which the good guys are shown to be bad and vice versa. The result is a displacement that undermines seemingly self-evident visions of what is good and what is bad, what is right and what is wrong, what is self and what is other.

In the original libretto, Osmin is a flat character, reduced to his distrust and the violent resolutions proposed to it; indeed, not a very charming image of a Turk. But if something is to be called grotesque here, it is not only the way Osmin presents an image of 'Turkish-ness' but also the image of Europeans which he presents to the audience. One might argue that Osmin's reduced and negative image of Europeans (as derived from his lines) characterizes him as stupid and short sighted, especially from the point of view of Europeans not wanting to recognize themselves in this image. Yet, one might also argue that visualizing him in a way that confirms that he is not to be taken seriously is a way of warding off an even bigger threat. For, the libretto proves him to be right. *Die Entführung aus dem Serail* shows Europeans to be treacherous, deceitful, without respect, and breaking of their promises. As the plot proceeds, Osmin's prejudices are shown to be to a large extent justified, and everything happens more or less as he originally predicted.

Within the context of the story, the Europeans' bad behaviour is justified as long as one accepts that Konstanza needs to be 'saved' from invasion by the Turk and that to this end all means are acceptable. But precisely this point of view is also what is questioned and destabilized by the opera itself, with the characterization of Osmin instrumental in causing such destabilization. His aggressive antipathy at first makes it easy to reject his way of looking at the Europeans as the product of a distorted vision, thus othering the image of the European self presented by Osmin, understanding the distortion as the vision of a stranger, someone who does not know. However, when this image in the end appears to be quite accurate, it is much harder to maintain the distinction between self and other.

This distinction is further problematized by the character of the other 'Turk', Bassa Selim. Selim trusts Belmonte, Pedrillo and Konstanze, gives them what they ask for, and in return they betray him. When he discovers how he has been deceived, he is furious and it seems he will use his power to take revenge, confirming the image of the violent Turk earlier presented by Osmin. The discovery that Belmonte is the son of the man who drove him from his belongings in Spain, threatened him and took his possessions, provides further justification for violent action. But then, not wanting to perpetuate such harmful patterns of behaviour, Bassa Selim decides to let them go. It is therefore not the Europeans that save Konstanze (as the title *Die Entführung aus dem Serail* might suggest) but rather Bassa Selim's contempt that sets them free, his refusal to be like them or even to have them near him any longer. They walk because he despises them.

Quraishi and Yalcin take this play with reversal of self and other one step further, reversing the situation in the plot. In their version, Belmonte, Pedrillo, Konstanze and her servant are young Turks, while Osmin and Bassa Selim represent old Europe. The story takes place not in Bassa

Selim's harem but in a theatre in Vienna or another old European city. With this strategy, they present a commentary not on how this opera shows the other but on what might be called a blind spot in the vision of self in which this opera invites us, the audience, to mirror ourselves. This blind spot is Konstanze's role, constructed around her honour that it is her responsibility to keep under all circumstances, although this honour is 'owned' not by her but by Belmonte, her fiancé. In the original libretto, right from the beginning, this honour, and not Konstanze's well-being is what is Belmonte's concern. Instead of being happy to see her again and being happy she is still alive, the first and only thing Belmonte is concerned about is if he isn't too late, if Bassa Selim has not taken her already, as if to be sure it still makes sense to rescue her. Furthermore, the anger his doubts about her honour evoke are directed against her, abducted against her will, and not against Bassa Selim. In reversing the roles, Quraishi exposes the uncanny similarities between Konstanze's role in this celebrated masterpiece of European culture and ways of dealing with women often criticized in others, this way questioning why we, the audience, would like to mirror ourselves in what is presented here.

In Qurashi's staging, it is the audience that 'steps inside' from the wings. Entering the theatre room over the main stage, they found themselves confronted with a square stage rising up in the middle of the auditorium. It is up to them to choose their own position, somewhere around the stage. Instead of being provided with a safe position in the dark from which to peer into an 'other' world through the *finestra aperta* of the proscenium arch, the relationship between seer and stage is made emphatically part of the performance.

On the raised stage, the action takes place through a series of poses, or tableaux vivants, rather than a continuous unfolding of dramatic action. This mise en scène highlights one of the structural characteristics that distinguish many operas from dramatic theatre, which is that the time structure of dramatic action is continuous whereas the time structure of opera is discontinuous. Arias, duets and choral sections expand on individual moments much beyond the limits of realistic representation and the action in between is often reduced to the bare minimum. The result is a structure that jumps from one intensified moment to the next.[7]

Quraishi's staging takes this structure to the extreme, reducing action to a series of poses that explicitly implicate the audience. The characters expose themselves to the look of audience, staging themselves as objects of their vision. The effect brings to mind Barthes's comparison of the tableau, as it functions in Diderot, to a fetish-object (Barthes, 1977, p. 71). Representation, Barthes argues, is not defined by imitation and therefore cannot be understood from the relation between the representation and the reality it is supposed to represent. Instead '[t]he 'Organon of Representation' ... will have as its dual foundation the sovereignty of the act of cutting out [decoupage] and the unity of the subject of that action' (Barthes, 1977, pp. 69–70). This duality, and not mimesis, is what constitutes representation. Here the tableau functions as the fetish-object where displacement seems to come to a halt in an image that can be seized by the eye.

For Barthes, Diderot is the theorist of the dialectics of desire as it is at work in representation. This desire is the subject of Quraishi's staging. He shows the characters and events, and by extension the Mozart opera, as fetish objects. The characters are perfectly stylized icons lighting up from the darkness surrounding them, as if cut out from their surrounding and put against a dark background. Idealized and perfectly self-contained, they present the promise of 'displacement coming to a halt'. At the same time, Quraishi's staging comments on these fetish characters, highlighting their construction as mirror images. The polished floor of the stage reflects the poses and tableaux, doubling the image and turning the opera into a house of mirrors in which the characters not only present ideal mirror images to the audience but are themselves always already reflections of other images that shape their appearances and through which they are seen (Figure 15). The performance complicates the relationship between the appearance of the characters and their reflection in the mirror-stage, between reality and representation. Bassa Selim is shown to be involved in a constant attempt to become the ideal body image in which he mirrors himself through body-building exercises, while Konstanze assumes the image of Ingres' *La Grande Odalisque*, a painting famous for its orientalist character, but also a reiteration of an older model in which the object is not the orientalist other but Venus. Ingres' oriental other mirrors the image of a European aesthetic model, conflating self and other in an ideal of female beauty and aesthetic composition. This oriental self-other in its turn becomes the image in which Konstanze's beauty reflects itself. With her perfectly styled looks and stylized behaviour, Konstanze assumes the image presented in the mirror of these paintings, thus confirming a cultural gaze in which self and other are mixed up. Performed by a Turkish opera singer, as it is the case in Quraishi's staging, Konstanze's appearance presents the reverse of Osmin and Bassa Selim staged as the oriental other in more conventional versions of *Die Entführung aus dem Serail*.

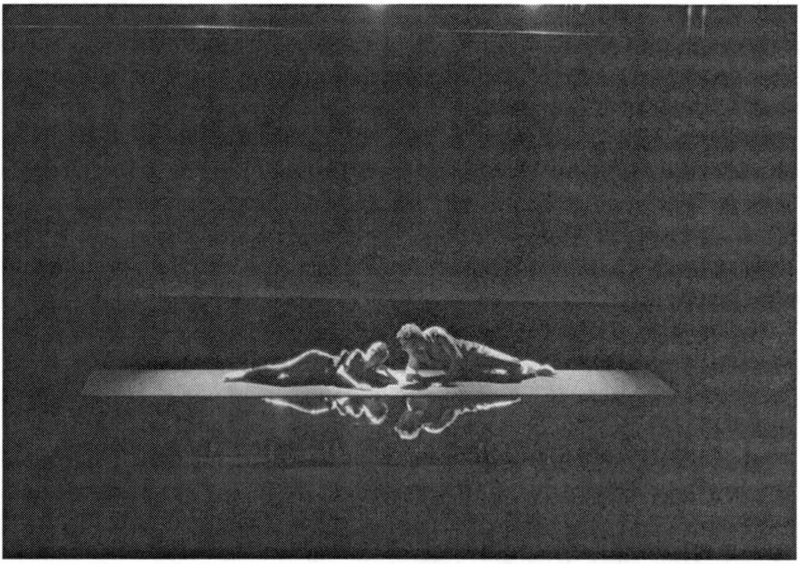

Figure 15 Photograph by Ronald Kiley showing Serap Gögüs and Ali Murat Erengül in Ibrahim Quraishi's *Saray. Mozart alla Turca*. Reproduced with permission of the photographer.

Mirroring herself in these paintings she becomes the mirror image presented to the audience, presenting this audience with an image in which to mirror itself, an image in which self and other are inextricably intertwined.

Notes

1 Dissecting Visuality in the Theatre – Introduction

1. *The Man Who (L'homme Qui).* A theatre performance by Peter Brook based on *The Man Who Mistook His Wife for a Hat* by Oliver Sacks. First production (*L'homme qui*) at Theatre des Bouffes du Nord, Paris, 1993. English version (The Man Who), 1994. With David Bennent, Sotigui Kouyaté, Bruce Myers, Yoshi Oïda and Mahmoud Tabrizi-Zadeh.
2. Oliver Sacks *The Man Who Mistook His Wife for a Hat* (London: Duckworth, 1985).
3. For a discussion of the terms vision, visuality and scopic regime and the way they are related, see Hal Foster's *Vision and Visuality*. Dia Art Foundation. Discussions in Contemporary Culture Number 2 (Seattle: Bay Press, 1988). In the introduction, Foster explains that although vision might suggest sight as a physical operation and visuality sight as a social fact, the two are not opposed as nature to culture:

   > [v]ision is social and historical too, and visuality involves the body and the psyche. Yet neither are they identical: here, the difference between the terms signals a difference within the visual – between the mechanism of sight and its historical techniques, between the datum of vision and its discursive determinations – a difference, many differences, among how we see, how we are able, allowed, or made to see, and how we see this seeing or the unseen therein. With its own rhetoric and representation, each scopic regime seeks to close out these differences: to make of its many social visualities one essential vision, or to order them in a natural hierarchy of sight (p. ix).

   'Scopic regime' is Christian Metz's term. See his *The Imaginary Signifier: Psychoanalysis and the Cinema*. Trans. Celia Britton et al. (Bloomington, IN: Indiana University Press, 1982) p. 61.
4. For an enlightening discussion of the question of the object of visual analysis, and the question of interdisciplinarity in relation to visual analysis, see Mieke Bal, 'Visual Essentialism and the Object of Visual Culture', *Journal of Visual Culture*, Vol. 2, No. 1 (2003) 5–32.
5. For two notable exceptions to this rule, see Freedman (1991) and Rokem (2003).
6. See also Silverman (1996).
7. See Sawday (1995) chapter 1, 'The Autoptic Vision'.
8. Since the beginning of the twentieth century, artists as diverse as Meyerhold, Brecht and Copeau have called for the retheatricalization of theatre as a place of artifice, and the re-establishment of theatrical reality as a more productive way of representing social life than naturalism. Retheatricalization aims at highlighting the rules and conventions of the stage. Brecht conceived of retheatricalization as a better way of depicting social reality, exposing its construction, and in this way denaturalizing it. See Patrice Pavis (1998) p. 395, Fischer-Lichte (1997) and Lehmann (1999).

9. My use of the term 'theoretical objects' is based on how this term was introduced to me by Mieke Bal in the ASCA Theory Seminar. See Bal (1999).
10. I take this idea of thought as movement from Deleuze and Guattari *What is Philosophy?* (New York: Columbia University Press, 1994).
11. In my text, I refer to Lehmann's book as it appeared in German in 1999 (*Postdramatisches Theater*). All translations are mine, with the original German in the footnotes. In 2006, Routledge published an English version of Lehmann's book (*Postdramatic Theatre*). This book contains the same argument, yet in an edited and abridged version.
12. The term *apparatus* originates from film theory, where it refers to the totality of interdependent operations that together make up the viewing situation. This includes(1) the technical base (the effects produced by the various components of the film equipment, including camera, lights, film and projection); (2) the conditions of film projection (dark theatre, immobility of spectators, the illuminated screen in front, and the light beam projected from behind the spectator's head); (3) the film itself as a 'text' (involving the various devices to represent visual continuity, the illusion of real space, and the creation of an illusion of reality); (4) the 'mental machinery' of the spectator (including conscious perceptual as well as unconscious and preconscious processes) that constitute the viewer as a subject of desire. The notion of the apparatus thus produces a definition of the entire cinema-machine that goes beyond films themselves and one that places the spectator – as unconscious desiring subject – at the centre of the entire process. See Robert Stam, Robert Borgoyne and Sandy Flitterman Lewis, eds, *New Vocabularies in Film Semiotics. Structuralism, Poststructuralism and Beyond* (London and New York: Routledge, 1992).

    The seminal texts in the theory of the apparatus are Jean-Louis Baudry's 'The Ideological Effects of the Basic Cinematic Apparatus' and 'The Apparatus: Metapsychological Approaches to the Impression of Reality in the Cinema', in Philip Rosen, ed., *Narrative, Apparatus, Ideology* (New York: Columbia University Press, 1986) pp. 286–318.
13. 'The logic of the Gaze is therefore subject to two great laws: the body (of the painter, of the viewer) is reduced to a single point, the *macula* of the retinal surface, and the moment of the Gaze (for the painter, for the viewer) is placed outside duration. Spatially and temporally, the act of viewing is constructed as the removal of the dimensions of space and time, as the disappearance of the body; the construction of an *acies mentis*, the punctual viewing subject' (Bryson, 1983, p. 96, italics in the original).
14. Bryson does acknowledge Gombrich's importance in rethinking painting from being a mere copy of reality towards a painting as a representation involving a subjective point of view. In *Art and Illusion*, Gombrich presents overwhelming evidence to show how the way we see and depict depends upon and varies with experience, practice, interest and attitudes. On the matter of perspective, however, Gombrich seems to take a position at odds with such relativity. Gombrich opposes the idea that perspective is merely a convention and does not represent the world as it looks (Gombrich, 1960, p. 254).

    See also Nelson Goodman *Languages of Art. An Approach to a Theory of Symbols* (London: Oxford University Press, 1969) p. 10.

15. 'Perspective', *The Oxford English Dictionary*. 2nd edn (Oxford: Clarendon Press, 1989).
16. This refers to Hubert Damisch's (1995) discussion of perspective. Damisch takes this idea of perspective as a symbolic form from Erwin Panofsky, who, in his turn, bases his notion of symbolic form on Ernst Cassirer.

    Erwin Panofski's article 'Die Perspektive als "symbolische Form"' was first published in *Vortrage des Bibliothek Warburg 1924–1925* (Leipzig, 1927) pp. 258–331. English translation by Christopher Wood, *Perspective as Symbolic Form*, New York, 1991. Ernst Cassirer *Philosophie der symbolischen Formen*. Vol. 1, *Die Sprache*, Berlin 1923. Vol. 2, *Der Mythos*, 1925. Vol. 3, *Phänomenologie der Erkenntnis*, 1929. English translation by Ralph Mannheim, *The Philosophy of Symbolic Forms*, Yale University Press. New Haven and London, 1955–1957. See Damisch (1995) chapter 1, (m Zone Books, New York, 1993).
17. 'Seer' *The Oxford English Dictionary*. 2nd edn (Oxford: Clarendon Press, 1989).
18. Deleuze called Foucault a 'seer' [voyant] in the sense that Foucault was seeing things that were not seen; Things that were invisible but not hidden. They are invisible in the visible. ('An interview with Gilles Deleuze', *History of the Present*, (Spring 1986), p 1. John Rajchman discusses Foucault as a seer in 'Foucault's Art of Seeing', in John Rajchman, *Philosophical Events of the '80s* (New York: Columbia University Press, 1991) pp. 68–102.

2 Step Inside!

1. Fried refers to Diderot's *Salons* (Jean Seznec and Jean Adhémar eds, *Salons I*, 2nd edn Oxford: Oxford University Press, 1975, *Salons II*, Oxford: Oxford University Press, 1960, *Salons III*, Oxford: Oxford University Press, 1963, and Jean Seznec ed., *Salons IV*, Oxford: 1977) and his *Oeuvres esthétiques* (ed., Paul Vernière, Paris: Garnier, 1966, includes *Entretiens sur le fils naturel, Discours de la poésie dramatique, Essais sur la peinture*, and *Pensées détachées sur la peinture*). My text is based on Fried's reading of these texts.
2. Fried's anti-theatrical prejudice is not directed against the theatre per se. In Fried's usage, the term 'theatricality' does not denote the essence, or even a quality of the theatre as an art form. Indeed, Fried mentions some theatre makers as being engaged in the same battle against theatricality as the modernist arts, citing Artaud and Brecht as examples. Instead, Fried uses theatricality to describe 'the wrong sort of consciousness of an audience', as he put it in 1987, in reviewing his 1967 essay (Fried, 1987, p. 57).

    Fried's definition of theatricality has given rise to rather different interpretations and uses, and has been opposed to various 'others'. Theatricality can serve as a figure for an emerging postmodernism, threatening to established modernism (Fried). But, theatricality can also be a figure representing a desiccated modernism against which an emergent postmodernism defines itself (Féral and Pontbriand). Theatricality became a polemical term both of condemnation (as exemplified by Fried's essay), and of praise (when used by supporters of postmodernism in the visual arts). For Fried, theatricality was the enemy of art, when art is understood from the perspective of Greenbergian modernism. Féral and Pontbriand argue, on very similar grounds, that theatricality is the enemy of art, when art is understood from a Derridean post-structuralist perspective. Discussing this amazing flexibility in the concept of theatricality, Auslander reminds his readers of Rosalind Krauss, who observes that 'theatre', in Fried's essay, is an empty term whose role it is to set up a system founded upon the opposition between itself and another term (Auslander, 1997, p. 52). Krauss calls this other term the 'nontheatrical' (Krauss, 1987, p. 62). Auslander adds that in Fried's historical account the nontheatrical is clearly modernism.

I agree with Krauss (and Auslander) that in Fried's essay theatricality is used to set up a binary opposition, with theatricality clearly the negative pole. However, it seems to me that what is empty is not theatricality but its opposite. Fried uses almost all of the 31 pages of his essay to explain what he means by theatricality. This explanation is not always consistent – it raises many questions and is open to different interpretations. It is not always clear whether theatricality is a quality of a work, or an effect produced in the interaction between a work and a beholder, or a sensibility (something either expressed in a work or by a beholder). Despite these confusions in Fried's conception of theatricality, if something has to be called empty, it is the opposite term – and this emptiness contributes to its status as the absolute. Only in his much later *Absorption and Theatricality* does this term get a name: absorption. Here, the opposition between theatricality and absorption functions as an act of discernment between *two* possible modes of relationship between painting and viewer, rather than theatricality as a relationship versus presentness as the absence of such a relationship.

This is a possibility already indicated in his 'Art and Objecthood', a possibility that is however pushed to the margin, in a footnote. In footnote 4, Fried takes issue with Greenberg's conception of modernism in painting as a progressive development, fuelled by an empirical search for the 'irreducible working essence of art and the separate arts'. This progression consists of seeking paintings' own formal essence through the rejection of the 'dispensable, unessential' conventions of its own tradition, as well as elements of the other arts. Although Fried adopts Greenberg's idea of modernism, he also attempts to embed the successive developments in painting within the historical moments that they appear. He writes:

> [T]he crucial question is not what these minimal and, so to speak, timeless conditions are, but rather what, at a given moment, is capable of compelling conviction, of succeeding as painting. This is not to say that painting *has no* essence; it is to claim that essence – i.e. that which compels conviction – is largely determined by, and therefore changes continually in response to, the vital work of the recent past. (Fried, 1968, pp. 123–124, italics in the orignal)

In this footnote, Fried rewrites Greenberg's timeless essences into effects on a viewer at a particular time and place, and turns the development of modern art into a continuous search for new strategies to *compel conviction*. Fried's account of the instantaneity of modern art thus puts the modernism he inherited from Greenberg into a temporal perspective, a perspective with which he does not seem to feel completely comfortable. Whereas he wants to make Greenberg's account of modernism more historical by building into it the idea that the essence of painting is historically contingent, he appears to be repulsed by art that he perceives as providing the viewer with an experience of such temporal contingency. This becomes the theme of his later work on the beginnings of the modern tradition in French painting.

3. In 'Art and Objecthood', Fried describes the presentness of modern art in terms that are remarkably similar to Norman Bryson's (1983) account of the 'presence' effect of perspective in painting. This analogy is all the more remarkable since modern painting, in Clement Greenberg's influential account of it, is characterized precisely by resistance to the illusionary effect of which perspective presents an image:

   > From Giotto to Courbet, the painter's first task had been to hollow out an illusion of three-dimensional space on a flat surface. One looked through this surface as through a proscenium into a stage. Modernism has rendered this stage shallower and shallower until now its backdrop has become the same as its curtain which has now become all that the painter has left to work on. (Greenberg, 1961, p. 136)

   Greenberg compares perspective in painting with the theatre and rejects both as illusory. In his account, the development of modern painting thus appears as a process of de-theatricalization of painting, analogous to the deconstruction of dramatic theatre.
4. This is not to say that dramatic theatre necessarily aims at obscuring the points of view implied by what is represented on stage. Freedman (1991), for example, demonstrates how Shakespeare's plays question rather than confirm the implications of subjective points of view as they are part of the dramatic representation.
5. *Artifact.* Choreography: William Forsythe. Music: Johan Sebastian Bach, Eva Crossman-Hecht. Stage, Lighting and Costumes: William Forsythe. Premiere: December 5, 1984, Ballett Frankfurt, Frankfurt.
6. J. L. Austin, *How to Do Things with Words. The William James Lectures* (Cambridge, MA: Harvard University Press, 1962); John Searle, *Speech Acts. An Essay in the Philosophy of Language* (Cambridge: Cambridge University Press, 1969). For a highly efficient introduction into speech-act theory and its subsequent developments, see chapter 7 of Jonathan Culler's *Literary Theory. A Very Short Introduction* (Oxford: Oxford University Press, 1997). For a discussion of speech-act theory in the context of theatre theory, see Carlson (1996) chapter 3, Elam (1988), Worthen (1998) and Andrew Parker and Eve Kosofski Sedgwick (eds), *Performance and Performativity* (New York and London: Routledge, 1995). See Foster (1998) for a critique of the notions of performance and performativity as derived from speech act theory. See Pavis (1998, pp. 280–284) for an overview of various pragmatic approaches to the theatre.
7. The term focalization was introduced by Gérard Genette to distinguish between two agents involved in the way events are represented in stories: the agent who 'narrates' and the agent who 'sees'. The concept was further developed by Mieke Bal upon whose work my use of the concept is based.

   Focalization originates from narratological theory designed for the analysis of verbal texts. Mieke Bal has demonstrated the utility of the concept for visual 'texts' as well. In Bal (1997), she uses a visual example – a relief in Arjuna, in southern India – to explain the principles of focalization. The relief shows the images of a cat, a man and several mice. Read correctly, these images 'tell' a story. Bal demonstrates how this visual story 'takes place' as a result of the spectator's ability to identify with the various positions presented by the mice, the cat and the man and to see from their respective points of view (Bal, 1997, pp. 144–145).

See Gérard Genette *Figures III*, Paris: Seuil 1972. Bal has demonstrated the usefulness of the concept of focalization for the analysis of both verbal and visual texts, in (among others) Bal (1991) and (1997). For an explanation of the theory of focalization, see especially 'Focalization' and 'Visual Stories' in Bal (1997) pp. 142–174, and Bal (2001).

8. Bruce Wilshire discusses the notion of 'standing in' in *Role Playing and Identity. The Limits of Theatre as a Metaphor* (1982). Wilshire introduces the notion of 'standing in' for the first time on pp. 22–23:

   The actor stands in for the character. But the character is a type of humanity with whom the audience member can identify, either directly as a stand-in for his own person, or indirectly as a stand-in for others whom the audience member recognizes and with whom he can be empathetically involved. If the character is one who stands in for us, then we can also stand in for him, and indeed we do stand in for him through the actor's standing in for him. ... actors stand in for characters who stand in for other characters. The audience member stands in through the actor's standing in for characters who stand in an actor like way, and if the audience member intuits a similarity to offstage existence, then this existence must be theatre like.

   See also chapter V: 'Variations on the Theatrical Theme of Standing in and Authorization.'
9. Bal refers to Louis Althusser *Lenin and Philosophy and Other Essays*. Trans. Ben Brewster (London: New Left Books, 1971).
10. See for example *Salons I* (64), *II* (197), *IV* (167, 359). As Fried points out, by Diderot's time the word *théâtral* had in addition to its meaning as pertaining to the theatre the pejorative one of a mode of action or expression which 'is suitable only for the theatre'. But it is only in Diderot's writings on drama and painting that the *maniéré* and the *théâtral* are in effect defined in terms of a positing of a beholder (see Fried, 1980, pp. 218–219).
11. I take the notion of 'gesture of exposure' from Bal (1996a) and discuss its use for the theatre more extensively in Chapter 7.

3 Showing What Cannot Be Seen

1. Although point of view coincides in terms of projection with the vanishing point, this does not mean that there is symmetry between them. As Damisch points out, strictly speaking vanishing point and point of view are situated, in three dimensional space, on a line perpendicular to the picture plane. But whereas the image of the point of view should be inscribed on the painting – at a virtual distance corresponding to that separating the spectator from the plane of projection – the vanishing point will be thrown far behind the image of the observer, who will have it, so to speak, at his back (Damisch, 1995, pp. 120–121). This ambiguity is crucial for the seductive appeal presented by perspective, an appeal to a desire for a stable and detached point of view outside what is seen that at the same time seems not to be outside at all. See also Chapter 9.
2. 'Das dramatische Theater, in dem die Bretter des Theaters die Welt bedeuten, konnte man mit der *Perspektive* vergleichen: der Raum ist hier im technischen wie im mentalen Sinn Fenster und Symbol, der Realität "dahinter"

analog. Er bietet ein sozusagen maßstäbliches, durch Abstraktion und Betonung gewonnennes metaphorisches Äquivalent der Welt wie das finestra aperta gedachte Renaissancegemälde' (Lehmann, 1999, p. 288; italics in the original).

3. 'Ob das Drama an verschiedenen Plätzen einer *Simultanbühne* spielt wie im Mittelalter, in der Mehrfachdekoration, dem "decor-multiple" der Renaissance, oder im typisierten Einheitsraum-Palast (palais à volonté) des Klassizismus, ob es vor dem Hintergrund des barocken "Schau- Platzes" für das Weltgeschehen oder im Kraftfeld eines naturalistischen Milieus stattfindet, das die Handlungen der Menschen vorab zu determinieren scheint, ist demgegenüber von untergeordneter Bedeutung: stets bleibt der dramatische Raum separiertes Symbol einer Welt als Totalität, sei diese auch noch so bruchstückhaft dargeboten' (Lehmann, 1999, p. 288; italics in the original).
4. 'Diese innere Ordnung, getragen von den berühmten Einheiten, schließt das Sinngebilde welches das Artefakt Tragödie darstellt, fugendicht nach außen gegen die Realität ab und konstituiert es zugleich im Inneren als lückenlose Einheit und Ganzheit' (Lehmann, 1999, p. 61).
5. This 'other world' shows no traces of an author: it appears as a self-contained unity. The Aristotelian unities, as they are part of the classicist tradition of dramatic writing, contribute to enforcing this illusion of a self-contained world.
6. This 'epic I' manifests itself in a variety of phenomena, such as an increasing tendency towards monologues at the cost of dramatic dialogue, a subjective perspective on events presented instead of the illusion of objective vision, direct address to an audience instead of the fourth wall and the undermining of the unities that guarantee the suggestion of a self-unfolding dramatic world.
7. See George Lakoff and Mark Johnson's argument in their *Metaphors We Live By* (1980).
8. I take this idea of the concept as a metaphor that replaces a story from Bal (1994). See also Chapter 1.
9. I use 'discourse' here to refer to the semiotic habits that enable us to communicate and think, and at the same time, prescribe ways of doing so. Discourse entails epistemological attitudes as well as unexamined assumptions about meaning and about the world. Language can be part of discourse but discourse is certainly not limited to language.
10. 'The history of this task over the centuries that followed bears a close resemblance to what Lorraine Daston calls the "history of objectivity"' (Fox Keller, 1994, p. 321). This is not a linear story, but a multi-layered and entangled one, accompanied by complex resistances and anxieties and by radical changes in the very meaning of the term objective. It is still, according to Fox Keller, possible to trace a distinctly linear arc in this non-linear story, a story line that is rooted in the very logic of scientific representation. This story line closely parallels the history traced by Rotman (1987), which is to say that this is a history of erasure, of the progressive disembodiment and dislocation of the scientific observer that ultimately became sufficiently complete to permit a comprehensive and apparently subject-less representation of the world.

    Daston points out that the term 'objective' had a very different – effectively opposite – meaning in the seventeenth century from what it has today: It referred neither to a state of mind, nor to a mode of perception, but to the objects of thought and perception, or to what Hobbes has called the 'effects of nature'. Only in the nineteenth century did the term 'objective' acquire the current meaning of a-perspectival – a 'view from nowhere', knowledge without a knower (Daston in Fox Keller, 1994, p. 315).

11. *The Children of Heracles* by Euripides. Ruhr-Triennale and Old Stories: New Lives. Iolaos: Jan Triska, Kopreus: Karen Kandel, Demophon: Brenda Wehle, Daughter of Heracles: Chris Chalk, Alkmene: Ruth Maleczech, Euristeus: Jan Triska. Premiere: 19 September 2002, Bottrop (Germany).
12. Zonneveld, 2004, p. 166.
13. *Antigona* by Tommaso Traetta. Muziektheater Transparant in collaboration with Brugge, Cultural Capital 2002 and Salamanca 2002. Director: Gerardjan Rijnders. Conductor: Paul Dombrecht. Orchestra: Il Fondamento. Chorus: La Sfera del Canto. Antigona: Rafaela Milanesi, Ismene: Giorgia Milanesi, Creonte: Guy de Mey, Emone: Maartje de Lint, Adastro: Markus Brutscher. Premiere: 7 September 2002, Concertgebouw Brugge.
14. *Penthesilea* by Heinrich von Kleist. Toneelgroep Amsterdam. Premiere: 22 October 1991, Stadsschouwburg Amsterdam.
15. *Andromache* by Jean Racine. Toneelgroep Amsterdam. Premiere 17 November 1990, Rotterdamse Schouwburg, Rotterdam.
16. *Klaagliederen,* based on the *Lamentations* by Jeremiah. Toneelgroep Amsterdam. Premiere: 16 November 1994, Stadsschouwburg Amsterdam.
17. See also my analysis of *Klaagliederen*, 'When God Looks Back: Rereading Religious Ritual on Stage', in Jonneke Bekkenkamp et al. eds, *Missing Links. Arts, Religion, Reality* (Münster-Hamburg-London: Lit. Verlag, 2000) pp. 31–48.
18. *Zinsbegoocheling* (*L'Illusion Comique*) by Pierre Corneille. Toneelgroep Amsterdam. Premiere: 16 October 1997, Stadsschouwburg Amsterdam.
19. *Dark Lady* (based on Shakespeare's sonnets). Toneelgroep Amsterdam and Muziektheater Amsterdam. Premiere: 9 January 1999, Muziektheater Amsterdam.
20. *Kwartett* by Heiner Müller. Toneelgroep Amsterdam. Premiere 13 December 1999, Stadsschouwburg Amsterdam.

4 Walking the Landscape Stage

1. *Sailors on a Bus* is a play by Dutch playwright and director Jeroen van den Berg. Premiere: 14 October, 1999, Grand Theatre, Groningen (The Netherlands). Translator of this quote is Michael Burke.
2. Hans Jonas 'The Nobility of Sight: A Study in the Phenomenology of the Senses', in *The Phenomenon of Life: Toward a Philosophical Biology* (Chicago: Northwestern University Press, 1982), quoted in Jay, 1993, p. 25.
3. *Picture Description/Explosion of a Memory*. An installation by Stefan Kunzmann. Text: Heiner Müller. Vienna: Hochschule für Angewandte Kunst, 1997.
4. 'Wo ist der Mensch' reads like a fundamental question of life. In the context in which it was asked, it can mean very simply 'Where is the actor or where is the space/place for the actor?' in the sense of where is the body/flesh mediating the text, the ideas, feelings etc.? But also: Where is the human being so central to the text of Heiner Müller and to the theatre or where is the agent?

5. 'While the architecture tends towards aleatoric devices and decomposition, the subject tends away from the centered Ego towards the murmuring voice of the unconscious (from meaning towards voice), and the body-voice loses its orientation by sense and meaning (direction voice-sound). On the whole, a fading of the pole of meaning takes place as well as a certain musicalization of the human voice tending towards sound patterns (Gertrude Stein). The line of the subject is weakened' (Lehmann, 1997, p. 59).
6. Chora refers to the pre-logical space that gives room to the play of being and becoming. Kristeva inherits the term 'chora' from Plato's account of the creation of the universe in his dialogue the Timaeus. Chora is part of her conception of signification as a dialectical process in which the semiotic manifests itself in the lusty disturbance of meaning, position and identity of subject and object as given in the symbolic structures that make up what is normally perceived as 'reality'.

   In his 1997 article, Lehmann speaks of a 'choreo-graphical inscription' (Lehmann, 1997, p. 57). In his later *Postdramatisches Theater* he speaks of theatre becoming 'Chora-graphie' (Lehmann, 1999, p. 263). In this later work, he contextualizes 'chora' primarily with reference to Julia Kristeva's *Revolution in Poetic Language* (New York: Columbia University Press, 1984). In his earlier (1997) text, Derrida has more prominence. Lehmann refers to Derrida's writing on the architectures of Bernard Tschumi and Peter Eisenmann, and to Derrida's term 'l'architecture de l'événement'. Derrida published a text titled 'Chora' in Jeffrey Kipnis and Thomas Leeser. *Chora L Works. Jacques Derrida and Peter Eisenmann* (New York: The Monacelli Press, 1997) pp. 15–32.
7. Conventional theatre, writes Stein, makes her nervous because it demands involvement with another time, which is the time of the drama characterized by accelerations and references backwards and forwards. With her notion of the landscape play, she argues for a theatre that simply happens here and now, and can be perceived like a landscape or a park.

   Lehmann (1997) p. 59 and (1999) p. 103 ff. See Gertrude Stein 'Plays' in Carl van Vechten, ed., *Last Operas and Plays by Gertrude Stein* (Baltimore and London: Johns Hopkins University Press, 1995). See also Bonnie Maranca *Ecologies of Theatre* (New York: Routledge, 1996) for an account of the influence of Gertrude Stein on the American avant-garde theatre and performance.
8. On the postmodern stage, the death of character manifests itself in a process that undermines the logocentric assumptions at work in conventional notions of theatrical presence. The illusion of spontaneous speech has been shattered. Where text appears as such, it is often distorted or transformed by means of textual strategies that aim at the deconstruction of fixed meaning and structural interruption.
9. See Jay (1993) chapter 9; '"Phallogocularcentrism": Derrida and Irigaray', pp. 493–542. With regards to the sun as metaphor, Jay refers to Derrida's 'The White Mythology: Metaphor in the Text of Philosophy', in *Margins of Philosophy*, Trans. Alan Bass (New York, London, Toronto: Harvester Wheatsheaf, 1982).
10. Barthes associates the experience of jouissance with what he terms 'writerly texts'. These are texts that, like the post-dramatic theatre, present an appeal to a reader to engage with and actively produce the text as what it is. In Barthes' account it is not the recipient recognizing him- or her-self in the

many 'others' presented by the text that causes the constitution of a new sense of selfhood at a distance. Rather the reverse: it is the impossibility of constituting unity that undermines the duality of the work as object versus the reader as subject. He opposes the writerly text to the readerly text, which is the type of text that supports the illusion that the unity of the text is something given rather than actively produced by a reader. The readerly text keeps intact the duality of the work as object versus the reader as subject. This duality is undermined by the writerly text. Fischer-Lichte similarly speaks of a decentred subject 'that ascertains its own identity by observing and becoming aware of just that decentering' (Fischer-Lichte, 1997, p. 59). For the distinction between readerly and writerly texts, see Roland Barthes, *S/Z* (New York: Hill and Wang, 1974). For the notion of jouissance, see Barthes (1975). This edition contains a foreword by Richard Howard in which he discusses the notions of pleasure and jouissance. Susan Foster (1986) uses Barthes' distinction between readerly and writerly, or between works and texts, as a starting point for a model for the analysis of contemporary dance as texts that afford the reader the opportunity to participate in the creation of meaning.

11. It is interesting in this respect that artist Stefan Kunzmann started the work on his 'staging' of Müller's text with an attempt at reconstructing the image from the text, but found it impossible to do so.
12. The painted landscape objectifies nature by framing it, by laying it before the viewer's eyes as a mastered entity. At the same time, the painted landscape is a sign that alludes to an infinitely larger presence outside the frame, one that cannot be encompassed by the eye. This sign can become a site for contemplative absorption, allowing an imaginary loss of the self in a larger order (Halkes, 2001, p. 8).
13. Seen this way, understanding the spectator as master of possible semiosis is indicative of a desire that also guides much (mis)use of the Derridean notion of dissemination as if meaning either unaccountability or the freedom to give meaning at will (see Chapter 5).

5 Navel Gazing as Critical Practice

1. *Bas and Elze Dance* (*Bas en Elze Dansen*), also written, directed and performed by Cas Enklaar and Els Ingeborg Smits. Premiere: January 18, 1996 in de Toneelschuur, Haarlem.
2. Perceived as such, the condition of the actor and his or her relation to the dramatic text could be called symptomatic for the general condition of being born into discourse and living one's life through the act of assuming positions within it. Bas and Elze have literally spent most of their lives expressing themselves on stage through the words of others. Even now, they often use the words of others to express their own fears and feelings. When, for example, they contemplate the time of their own death approaching, they use the text of Chekhov's *Three Sisters*:

   The years will pass, and we shall all be gone for good and quite forgotten. ... Our faces and our voices will be forgotten and people won't even know that there were once three of us here. ... But our sufferings may mean happiness for the people who come after us. ... There will be a time when

> peace and happiness reign in the world, and then we shall be remembered kindly and blessed. ... Oh, if only we knew, if only we knew (Chekhov, 1954, pp. 329–330).

When, a little later, Elze lapses into a momentarily state of absence, Bas contemplates the fragility of life, repeating King Lear's speech when he discovers his daughter Cordelia's dead body.

3. Cor van der Lugt-Melsert (1882–1969) actor, director and manager of theatre company Het Hofstad Tooneel (The Hague) and of the city theatre of Amsterdam. After World War II, he was suspended from his office for a period of two years. He began working as a theatre critic and in 1949 he published his autobiography *Wat ik nog zeggen wou ...* , (Maastricht: Leiter-Nypels, 1949).
4. Original in Dutch: 'Niet meer het technisch kunnen om "in andermans huid" te kruipen, maar de persoonlijkheid van de acteur als mens komt op de voorgrond. De invloed van dans en performance versterkten nog deze tendens tot het verdwijnen van het personage' (Van Kerkhoven, 1998, p. 111).
5. Peter Brook *The Empty Space* (London: Macgibbon & Kee, 1968)
6. Cas Enklaar and Els Ingeborg Smits (the actors performing Bas and Elze), although not old enough to be called fossils themselves, are the product of the theatrical developments their performance is about. These developments have defined their 'role' as actors. Their careers started, respectively, in 1968 and 1969, a time of revolt in Dutch theatre that has gone down in history as the Aktie Tomaat (Action Tomato). On 9 October 1969, two students of the Amsterdam Theatre School threw tomatoes at the stage at the end of a performance by the Nederlandse Comedie of Shakespeare's *The Tempest*. Their action was the start of a series of protest-manifestations with which young theatre makers revolted against what they saw as uninspired and old fashioned repertoire theatre, as well as against the old fashioned, hierarchically organized companies that dominated the existing order. This revolt inaugurated major changes in Dutch theatre establishment and ended the careers of many leading actors and actresses who had made it in the more conventional theatre. The role of Elze is based on the memoirs of several such actresses. Actors Cas Enklaar and Els Ingeborg Smits themselves belong to the generation that threw tomatoes and became famous for resisting the type of theatre represented by the characters Bas and Elze.
7. Barthes writes: 'To recognize the studium is inevitably to encounter the photographer's intentions, to enter into harmony with them, to approve or disapprove of them, but always to understand them' (Barthes, 1993, pp. 27–28). And: 'Certain details may "prick" me. If they do not, it is doubtless because the photographer has put them there intentionally' (p. 47). 'The studium is ultimately always coded, the punctum is not' (p. 51).
8. See Bal (1991, pp. 18–24) and (2001, pp. 65–91).
9. Bal:

> between the text (the story of the welcomed arrival of Zeus) and the image (the exhibition of a female body for voyeuristic consumption), the painting produces its own narrative, reducible to neither – the work's visual/narrative textuality. The pre-text is literally a pretext: Its anteriority allows the painting's appeal to the general story as a frame for its reversal. The story's centrality, as the theme of the work, allows everything decentered to slip in: it allows, that is, for the dissemination of meaning. ...

> These two stories – the purely textual, verbal pre-text and the story of the purely visual present – collude and collide in the work's textuality. They are in tension, but not in contradiction. They produce a new story, the text of the Danae as an interaction between the canvas and the viewer who processes it. In this text, Zeus, invisible as he/it is, thus becomes the pre-text the woman uses to get rid of the indiscreet viewer. The woman who at first sight seemed to be on display – as a spectacle, in a static visual reading – takes over and dominates both viewer and lover. Her genitals, prefigured by the slippers and magnified by the opening of the curtain at the other end of the sight line, are central in the framed text. They are turned toward the viewer, but they can be seen by neither viewer nor lover, because the viewer is sent away while the lover comes to her from the other side/sight. This way, her sexuality, in spite of its centrality, is a trace of the pre-text, for the conflicting lines of sight cut it off; it is also the locus of the metaphor that kept creeping into the vocabulary of my analysis: It is the navel of the text. But between sex and navel lies a difference – the difference between voyeurism and its deconstruction. (Bal, 1991, p. 21)

6 Retheatricalizing Sexuality in the Field of Vision

1. This third term involved in the way our culturally determined look 'takes place' is crucial to Silverman's rereading of Lacan and will be discussed at length in the following chapter.
2. For a discussion of Annie Sprinkle's *Post Porn Modernist* in relation to *Etant donnés* and *L'Origine du Monde*, see Rebecca Schneider *The Explicit Body in Performance* (London and New York: Routledge 1997) chapter 2, 'Logic of the Twister, Eye of the Storm'.
3. *Looking for peter* Choreography: Gonnie Heggen. Dancers: Gonnie Heggen, Yasuko Yokoshi, Noortje Bijvoets, Vivianne Rodrigues de Brito. Peter Pan: Nini Poortman. Premiere: 14 November 1996, Theater Frascati, Amsterdam.
4. See Halkes (2001), chapter 1. The term 'recovery narrative' is from Caroline Merchant 'Reinventing Eden: Western Culture as Recovery Narrative', ed., William Cronon *Uncommon Ground: Towards Reinventing Nature* (New York and London: W. W. North and Company, 1995).
5. See Sigmund Freud, 'Some Psychical Consequences of the Anatomical Distinction Between the Sexes', in *The Standard Edition of the Complete Psychological Works* (London: Hogart, 1961), Vol. 21, pp. 149–157, and Sigmund Freud, 'Female Sexuality', in *The Standard Edition*, Vol. 21, pp. 225–243 and Sigmund Freud, 'Femininity', in *The Standard Edition*, Vol. 22, pp. 112–135.
6. 'The Signification of the Phallus' in Lacan, 1977, pp. 281–291. See also Grosz (1990).
7. Original in Dutch: 'een vrolijk esoterisch boswezentje dat van alle godsdiensten een hapje heeft genomen en niets heeft doorgeslikt zodat het onbekommerd zijn dwaallicht kan laten schijnen over alles wat groeit en bloeit' Ariejan Korteweg, *De Volkskrant*, 19 November 1996.
8. Silverman points out that, within Lacan's own texts, the signifier phallus sustains two different meanings, neither of which entirely maintains its

autonomy from the penis. On the one hand, the phallus appears as the signifier of 'the fullness of being', of those things which have been partitioned off from the subject during various stages of its constitution, and which will never be restored to it. On the other hand, the phallus is a signifier for the cultural privileges and positive values which define male subjectivity within patriarchal society, but from which the female subject remains isolated (Silverman, 1983, pp. 183–184).

9. 'The Lacanian's desire clearly to separate *phallus* from *penis*, to control the meaning of the signifier *phallus*, is precisely symptomatic of their desire to have the phallus, that is, the desire to be at the center of language, at its origin. And their inability to control the meaning of the word *phallus* is evidence of what Lacan calls symbolic castration' (Gallop, 1988, p. 126; italics in the original).

7 Disorders That Consciousness Can Produce

1. *De Zieleweg van de Danser* (*The Path of the Dancer's Soul*) was first presented at the International Theaterschool Festival 1997 in Amsterdam. Director: Gerardjan Rijnders. Choreography: Krisztina de Châtel. Performed by: Mimoun Oaïssa and Wen-Cheng Lee.

   The Path of the Dancer's Soul blurs the boundary between dance and text-based theatre, a move that, as pointed out in the previous chapters, is typical of many experiments by theatre makers in the 1980s and 1990s, including Gerardjan Rijnders himself. Rijnders worked with the Dutch National Ballet for his version of *The Bacchae* (1986), and later used the experiences he gained from working with these dancers to create *Ballet* (1990) with actors from his own company. *The Path of the Dancer's Soul* is a continuation of these explorations.
2. This first version of the mirror stage was never published. The text that opens Alan Sheridan's collection of Lacan's texts in *Écrits* is a translation of a later version of the text, presented at the International Congress of Psychoanalysis in Zürich in 1949. For a detailed account on the confusion around the correct date of Lacan's text, see Jane Gallop 'Where to begin?' in *Reading Lacan* (Ithaca and London: Cornell University Press, 1985) pp. 74–92.
3. Silverman refers to Paul Schilder's *The Image and Appearance of the Human Body: Studies in the Constructive Energies of the Psyche* (New York: International Universities Press, 1950); Henri Wallon. *Les origines du caractère chez l'enfant: les préludes du sentiment de personnalité* (Paris: Boivin et Cie, 1934); Henri Wallon. 'Kinesthesia and the Visual Body Image in the Child', in Gilbert Voyat, ed., *The World of Henri Wallon* (New York: Jason Aronson, 1984); and an essay by Maurice Merleau Ponty on Wallon: *Les Relations avec autrui chez l'enfant* (Paris: Centre de Documentation Universitaire, 1958).
4. 'Assailed at various points, [my pre-existing] corporeal schema crumble[s]', he recounts. 'I [subject] myself to an objective examination, I [discover] my blackness, my ethnic characteristics; and I [am] battered down by tom-toms, cannibalism, intellectual deficiency, fetichism [sic], racial defects, slave ships, and above all: "Sho' good eatin"' (Fanon, quoted in Silverman, 1996, p. 28).
5. Jean-Paul Sartre *Being and Nothingness. An Essay on Phenomenological Ontology*. Trans. Hazel E. Barnes (London: Methuen, 1959). Original in French *L'être et le néant*, 1953.

8 Death, Digitalization and Dys-Appearance

1. In his mirror stage essay Lacan refers to how the fragmented body can manifest itself in dreams. It then appears in the form of disjointed limbs, or of in those organs represented in exoscopy, growing wings and taking up arms for intestinal persecutions – the very same that the visionary Hieronymus Bosch has fixed, for all time, in painting, in their ascent from the fifteenth-century to the imaginary zenith of modern man (Lacan, 1977, pp. 4–5). Lacan comes back to this theme in his 'Some Reflections on the Ego', *International Journal of Psychoanalysis*, Vol. 34, 1953. See also Grosz (1994, pp. 39–46).
2. A first version of *Holoman: Digital Cadaver* was presented at the Festival a/d Werf (Utrecht, May 1997). In this text, I refer to the second version of the show performed in De Balie, Amsterdam, 26 and 27 May 1998. Performed by Frank Sheppard (Holoman). Concept/director/music (electronics, bass): Mike Tyler. Guitars: Meindert Meindertsma. Percussion: Peter Kuitwaard. Choreography: Frank van de Ven. Animations, video: Isabelle Jenniches.

   For more information and images, see: http://www.media-gn.nl/mfa/isabelle/HOLOMAN
3. The name of the person used to produce these images was not officially disclosed. However, since the date of his execution was released it has not been very difficult for journalists to discover that the person in question was Joseph Paul Jernigan, found guilty of robbing and murder and executed on 26 August 1993.
4. This quote is taken from the The Visible Human Fact Sheet http://www.nlm.nih.gov/pubs/factsheets/visible_human.html
5. For more information on the technical aspects of the Visible Human Project see: http://www.dlib.org/dlib/october95/10ackerman.htlm

   See also the website of the Visible Human Project: http://www.nlm.nih.gov/research/visible
6. The Visible Human Female data set was released in November 1995 and has the same characteristics as the human male, with one exception. Due to new techniques, it was possible to cut the female cadaver into even thinner slices. This time no less than 39 gigabite of information found its way to cyberspace.
7. Inscriptions in Dutch: 'Zij die bij hun leven als boosdoener schaadden, worden van nut na hun dood en gezondheid ontleent aan den dood zelf haar bevordering.'

   'Voorhoofd, vinger, nier, tong, hoofd, long, hersens, handen, geven u levenden waarschuwend voorbeeld.'

   'Hoorder neemt u ter harte, en terwijl u gaat langs de verschillende dingen van het leven. Weest er van overtuigd dat ook in het nietigste de Godheid nog verborgen is.'
8. These similarities between the Visible Human Project and the historical anatomical dissection were pointed out by José van Dijck in a lecture presented in De Balie in Amsterdam (18 December 1997). See also my review of this lecture in relation to the performance Holoman: Digital Cadaver; 'Het Visible Human Project als Schouwtoneel van de 21ste Eeuw', *Etcetera XIV* (64) 1998, 19–23. See van Dijck (2000).

   See also Sawday (1995), Thacker (1998), Ruth Richardson *Death, Dissection and the Destitute* (London and New York: Routledge and Kegan Paul, 1987) and

Thomas Tierney 'Anatomy and Governmentality: A Foucauldian Perspective on Death and Medicine in Modernity', *Theory and Event*, Vol. 2, no. 1 (1998) 1–31. The relationship between anatomy and theatre in past and present is subject of my *Anatomy Live. Performance and the Operating Theatre.* (Amsterdam: Amsterdam University Press, 2008).

9 Managing the Attention of the Audience

1. Typical in this respect is that Silverman first discusses Paul Schilder (*The Image and Appearance of the Human Body*, 1950), who understands the body image as the effect of synaesthetic processes in which various sensations are brought together, but then turns to Wallon for a model in which the visual imago or 'exteroceptive ego' is initially disjunctive from what he calls the 'proprioceptive ego'. According to Schilder, the sensational and visual dimensions of the bodily ego are so closely integrated with one another as to give rise to a unified sense of self. In his account, the body image describes the way the body appears to ourselves and social and interpersonal attachments and investments, as well as libidinal energy, form a part of this self-image and conception. The body image is formed out of the various modes of contact the subject has with its environment through its actions in the world and through the actions of others. Optical or visual aspects are part of this mental representation of the body, as well as tactile, kinaesthetic, cutaneous and proprioceptive sensations. For Schilder, every touch is already oriented in a visual register. However, he also stresses that the body schema does not have two parts, one optical and the other tactile. The body image is synaesthetic, just as every sensation is in fact synaesthetically organized and represented. He writes 'This means that there does not exist any primary isolation between the different senses. The isolation is secondary' (Schilder, 1950, p. 38).
2. In the Panopticon, the guard represents a 'function of seeingness' (Lacan) similar to the Lacanian gaze, a function that indeed might be called a manifestation of the symbolic in the field of vision, as Silverman, discussing Lacan, so eloquently puts it. As she demonstrates, it is not only because the gaze illuminates how culture is at work in the kind of body images that mediate in both our sense of self as a body seen and how we see. The Lacanian gaze explains how even that which falls outside our field of vision guides how we see what we see, and thus confronts us with the limitations of our subjective perspective.
3. *Live*. Choreografie: Hans van Manen. Performed by Coleen Davis and Henny Jurriens (dance), Henk van Dijk (Camera), and Ed Spanjaard (Piano). Het Nationale Ballet. Premiere: 2 June 1979, Theater Carré, Amsterdam.
4. *Double Track*. Choreography and concept: Beppie Blankert. Music: Louis Andriessen. Text: Samuel Beckett. Dancers: Beppie Blankert and Caroline Dokter. Sticthing Dansproductie. Premiere: 21 October 1986, Mickery, Amsterdam.

   Double Track, een remake. Choreography and concept: Beppie Blankert. Music: Louis Andriessen. Text: Samuel Beckett. Dancers: John Taylor and Christopher Steel. Blankert Dansconcerten and Stichting Grand Theatre. Premiere: 17 November 1999, Theater aan het Spui, Den Haag.

5. Lacan refers to Wallon in 'Some Reflections on the Ego' in *International Journal of Psychoanalysis*, Vol. 34 (1935), Lacan (1977, p. 3) and Lacan (1981, pp. 73, 99, 109). See also Grosz (1990, pp. 35–38).
6. Lacan refers to the behaviour of pigs and migratory locusts, as well as to Callois legendary psychasthenia to explain this behaviour as a way of installing a relationship between *Innenwelt* and *Umwelt*, just as in the mirror stage (Lacan, 1977, pp. 3–4). He comes back to this notion of mimicry in his later *Four Fundamental Concepts in Psychoanalysis*, where he argues against the idea that mimicry in certain animals serves as a protective device. Instead, he argues, mimicry is part of a strategy to become part of a particular 'picture' (Lacan, 1981, pp. 73, 98–100, 107, 109).
7. Lacan's recourse to the metaphor of a stain when accounting for the image in the guise of which we invite the gaze to affirm us suggests the need for a more supple understanding of the relation between our bodies and the representations which make up the cultural inventory suggested by the signifier 'screen'. The stain metaphor accounts for that relation in three-dimensional rather than two-dimensional terms, and it collapses the distance between the body and the image which defines it. ... It designates, in a way the screen cannot, the transformation of actual muscles and flesh into a photographic representation, and it helps to understand that this representation can implicate the postural schema and indeed the whole of what Wallon calls 'proprioceptivity'. It can thus involve a corporeal assimilation of the image. (Silverman, 1996, pp. 201–202)

   Mimicry does not always imply a resistant or even a conscious intentionality: on the contrary, it may bespeak a subject's completely unconscious compliance with the images to which he or she is accustomed to being apprehended by the camera/gaze. The pose needs to be more generally understood as the photographic imprinting of the body, and that imprinting is not always apparent to the subject in question. It may be the result of the projection of a particular image onto the body so repeatedly as to induce both psychic and corporeal identification with it. And the image may be generative not of pleasure, but unpleasure. (Silverman, 1996, p. 205)
8. [M]any circuits of sense impression and movement have become so familiar with repetition that they have made well-worn paths for themselves in the neuromuscular system and now operate without our even being aware of them. A circuit that is often traversed becomes, as in any other travel route, increasingly familiar with each repetition until it is virtually automatic. (Martin, 1939, p. 43)
9. These movement-sense receptors comprise certainly the busiest system in the entire body, for we are in a continual state of postural change, far greater than we realize, from the movement of the eyeballs in following an object, to the periodic contraction of the stomach. Besides the receptors in the muscle tissue, there are also semicircular canals in the ears which collaborate in recording data concerning our balance and tendencies to shift from it. If we did not have this testimony, we should be perpetually falling down, or even unaware of whether we were right side up or upside down. Through the agency of movement sense, we are able to regulate the force of our movements, to co-ordinate them so that objects can be picked up or put down, and to make

hundreds more otherwise impossible motor adjustments without which we could not begin to carry on a single day's normal activity. (Martin, 1939, p. 44)

10. Movement sense in Martin's account is proprioceptive in the sense that it is a function of our proprioceptive systems. Proprioceptive systems are those channels of information whose source is the body. These systems provide information about the body, but they are not the only systems that do so. For example, information about movement and relative posture of the body is also available in visual flow, to which the vestibular system is sensitive (as well as being sensitive to gravity-relative movement and posture). So, it is neither true that internal proprioceptive systems can provide information only about the body. Proprioceptive systems can also contribute to our awareness of the world this body is confronted with. Nor is it true that information about the body comes only via the internal proprioceptive systems. Finally, not all this information results in proprioceptive awareness, where this is taken to be conscious experience of the body characterized as experience of the body from the inside. Few of these different types of information are consciously registered and when they do generate conscious experiences the contents of the experience are often different from the contents of the information that generates the experience. Often, no clear distinction is being made between these three aspects of proprioception (systems, information and awareness), which is the cause of much confusion. See for an extensive discussion of this distinction, 'Self-Consciousness and the Body: An Interdisciplinary Introduction', in José Luis Bermúdez, Anthony Marcel and Naomi Eilan, eds, *The Body and the Self* (Cambridge, MA: The MIT Press, 1995).
11. With his notion of the senses as perceptual systems, Gibson objects to the model of a receptor mosaic for each sense connecting with the central nervous system and projecting the pattern of excited receptors in the brain. He argues that, instead of supposing that the brain constructs or computes the objective information from a kaleidoscopic inflow of sensations, we may suppose that the orientation of the organs of perception is governed by the brain so that the whole system of input and output resonates with the external information. If this formula is correct, the input of the sensory nerves is not the basis for perception but only half of it. It is only the basis for sense impressions. These are not the data of perception, nor the raw material out of which perception is fashioned in the brain. The active senses cannot be simply the initiators of signals in the nerve fibres or messages to the brain, instead they are analogous to tentacles and feelers. The function of the brain when looped with its perceptual organs is not to decode signals or to interpret messages or to accept images. The function of the brain is not even to organize sensory input or to process data. Perceptual systems, including the nerve centres at various levels up to the brain, are ways of seeking and extracting information about the environment from the flowering array of ambient energy.

10 Welcome to What You Think You See

1. See Chapter 2, note 4.
2. See (among others), John W. van Cleve *Humanität und Kreuzzugsideologie um 1780. Die 'Türkenoper' im 18. Jahrhundert und das rettungsmotiv in Wielands*

'Oberon', Lessings 'Nathan' und Goethe's 'Iphigenie', Kanadische Studien zur deutsche Sprache und Literatur, 30 (Bern/New York: Lang, 18984); Timothy D. Taylor 'Peopling the Stage: Opera, Otherness, and New Musical Representations in the Eighteenth Century', *Cultural Critique* (Spring 1997) 55–88; Eve Meyer 'Turquerie and Eighteenth-Century Music', *Eighteenth Century Studies*, Vol. 7, no. 4 (Summer 1974) 474–488.

3. Konstanze: Malin Hartelius, Blonde: Patricia Petitbon, Belmonte: Piotr Beczala, Pedrillo: Boguslaw Bidzinski, Osmin: Alfred Muff, Bassa Selim: Klaus Maria Brandauer. Orchestra and chorus of the Opernhaus Zürich. Conductor: Christoph König. Director: Jonathan Miller. Premiere: 22 June 2003, Opernhaus Zurich. DVD: Bel Air Classiques.
4. Konstanze: Eva Mei, Blode: Patrizia Ciofi, Belmonte: Rainer Trost, Pedrillo: Mehrzad Montazeri, Osmin: Kurt Rydl, Bassa Selim: Markus John. Orchestra and Chorus of the Maggio Musicale Fiorentino. Conductor: Zubin Metha. Director: Massimo Teoldi. Teatro della Pergola, Florence, 2002. DVD: Rai Trade.
5. Konstanze: Els Dottermans & Rena Granieri, Belmonte: Jan Decleir & Christian Baumgärtel, Blondje: Annet Malherbe & Selma Harkink, Pedrillo: Tom de Wispelaere & Peter Gijsbertsen, Osmin: Stefaan Degand, Bassa Selim: Najib Cherradi. Orchestra: Beethoven Academie. Musical adaptation and composition: Wim Henderickx. Conductor: Koen Kessels. Adaptation and director: Ramsey Nasr. Premiere: 13 September 2006, DeSingel, Antwerpen.
6. With Serap Gögüs, Alev Irmak, Cigdem Soyarslan, Görkem Ezgi Yildirim, Michael Doumas, Erdem Erdogan, Ali Murat Erengül, Martin Niedermair. Orchestra: Serdar Yalcin, Martin August Fuchsberger, Didem Basar Dermen, Serkan Mesut Halili, Güniz Yilmaz, Binnaz Celik, Neva Özgen, Eren Özek, Tugut Aktas. Musical adaptation, composition and conductor: Serdar Yalcin. Adaptation: Ibrahim Quraishi and Gabriel Smeets. Director: Ibrahim Quraishi. Premiere: 19 September 2006, Schauspielhaus Vienna.
7. See for an extensive discussion of the implications of this structure: Carl Dahlhaus. *From Musikdrama zur Literaturoper. Aufsätze zur neueren Operngeschichte*. München/Salzburg: Musikverlag Emil Katzbichler, 1983.

Bibliography

Auslander, Philip, 1997, *From Acting to Performance* (London and New York: Routledge).

Bal, Mieke, 1991, *Reading 'Rembrandt': Beyond the Word-Image Opposition* (New York: Cambridge University Press).

——, 1994, 'Scared to Death' in *The Point of Theory. Practices of Cultural Analysis*, eds, Mieke Bal and Inge Boer (Amsterdam: Amsterdam University Press) pp. 32–47.

——, 1996a, *Double Exposures. The Subject of Cultural Analysis* (New York and London: Routledge).

——, 1996b, 'Reading Art', in *Generations and Geographies in the Visual Arts – Feminist Readings*, ed., Griselda Pollock (London and New York: Routledge) pp. 25–40.

——, 1997, *Narratology. Introduction to the Theory of Narrative*, 2nd edn (Toronto, Buffalo, London: University of Toronto Press).

——, 1999, *Quoting Caravaggio: Contemporary Art, Preposterous History* (Chicago and London: The University of Chicago Press).

——, 2001, *Looking In: the Art of Viewing* (Amsterdam: G + B Arts International).

Barthes, Roland, 1975, *The Pleasure of the Text*. Trans. Richard Miller (New York: Hill and Wang).

——, 1977, *Image, Music, Text*. Essays selected and translated by Stephen Heath. (London: Fontana Press).

——, 1993, *Camera Lucida. Reflections on Photography*. Trans. Richard Howard. (London: Vintage Books).

Beckett, Samuel, 1974, *Texts for Nothing* (London: Calder and Boyars).

——, 1990, *The Complete Dramatic Works* (London and Boston: Faber and Faber).

Benveniste, Emile, 1971, *Problems in General Linguistics*. Trans. Mary Elisabeth Meek (Coral Gables, FL: University of Miami Press).

Bolla, Peter de, 1995, 'The Visibility of Visuality: Vauxhall Gardens and the Siting of the Viewer' in *Vision and Textuality*, eds, Stephen Melville and Bill Readings (Basingstoke: Macmilan) pp. 282–295.

Boulter, Jonathan, 1999, '"Wordshit, Bury Me": The Waste of Narrative in Samuel Beckett's "Texts for Nothing"', Lecture presented at the conference 'The Capital of Ruins: Beckett's Prose Fragments' 3 June, South Bank University, London.

Bronfen, Elisabeth, 1994, 'Death, the Navel of the Image', in *The Point of Theory. Practices of Cultural Analysis*, eds, Mieke Bal and Inge Boer (Amsterdam: Amsterdam University Press) pp. 79–90.

Bruner, Jerome, 1990, *Acts of Meaning* (Cambridge, MA and London: Harvard University Press).

Bryson, Norman, 1983, *Vision and Painting. The Logic of the Gaze* (London: Macmillian).

Buck-Morss, Susan, 1992, 'Aesthetics and Anaesthetics: Walter Benjamin's Artwork Essay Reconsidered', *October*, 62 (Fall) 3–40.

Buck-Morss, Susan, 2003, *Thinking Past Terror. Islamism and Critical Theory on the Left* (London and New York: Verso).
Butler, Judith, 1990, *Gender Trouble* (New York and London: Routledge).
——, 1993, *Bodies that Matter: On the Discursive Limits of Sex* (New York and London: Routledge).
Carlson, Marvin, 1993, *Theories of the Theatre: A Historical and Critical Survey, from the Greeks to the Present*, expanded edition (Ithaca and London: Cornell University Press).
——, 1996, *Performance: A Critical Introduction* (London and New York: Routledge).
Case, Sue Ellen, 1995, 'Performing Lesbian in the Space of Technology: Part II', *Theatre Journal*, 47, 329–343.
Checkhov, Anton, 1954, *Plays*. Translated and with an introduction by Elisaveta Fen (London: Penguin Books).
Crary, Jonanthan, 1992, *Techniques of the Observer. On Vision and Modernity in the Nineteenth Century* (Cambridge, MA and London: MIT Press).
——, 1999, *Suspensions of Perception. Attention, Spectacle and Modern Culture* (Cambridge, MA and London: MIT Press).
Damisch, Hubert, 1995, *The Origin of Perspective*. Trans. John Goodman (Cambridge, MA and London: MIT Press).
Derrida, Jacques, 1993, *Limited Inc* (Evanston, IL: Northwestern University Press).
Descartes, René, 1977, *René Descartes. The Essential Writings*. Translated and with an introduction and a conceptual index by John J. Blom (New York: Harper and Row).
Diamond, Elin, 1996, 'Introduction', in *Performance and Cultural Politics*, ed., Elin Diamond (London and New York: Routledge) pp. 1–13.
Dijck, José van, 2000, 'Digital Cadavers: The Visible Human Project as Anatomical Theater', in *Studies in History, Philosophy, Biology and Biomedical Science*, 31 (2), 271–285.
Elam, Keir, 1980, *The Semiotics of Theatre and Drama* (London and New York: Routledge).
——, 1988, 'Much Ado About Doing Things With Words (and Other Means): Some Problems in the Pragmatics of Theatre and Drama', in *Performing Texts*, eds, M. Issacharoff and R. F. Jones (Philadelphia: University of Pennsylvania Press) pp. 39–58.
Elkins, James, 1996, *The Object Stares Back. On the Nature of Seeing* (San Diego, New York and London: Harcourt Brace & Company).
Euripides, 1981, *The Children of Heracles*, trans. Henry Taylor and Robert A. Brooks (New York and Oxford: Oxford University Press).
Faunce, Sarah and Linda Nochlin, 1988, *Courbet Reconsidered* (Brooklyn, New York: The Brooklyn Museum).
Féral, Josette, 1982, 'Performance and Theatricality: The Subject Demystified', *Modern Drama*, XXV, 170–180.
Fischer-Lichte, Erika, 1992, *The Semiotics of Theatre*. Trans. Jeremy Gaines and Doris L. Jones (Bloomington and Indianapolis: Indiana University Press).
——, 1997, *The Show and the Gaze of Theatre: A European Perspective* (Iowa City: University of Iowa Press).
Foster, Susan Leigh, 1986, *Reading Dancing: Bodies and Subjects in Contemporary American Dance* (Berkeley, Los Angeles, London: University of California Press).

——, 1998, 'Choreographies of Gender', *Signs: Journal of Women in Culture and Society*, 24 (1), 1–33.
——, 'Kinesthetic Empathies and the Politics of Compassion', unpublished conference paper.
Foucault, Michel, 1980, *Power/Knowledge: Selected Interviews and Other Writings 1972–1977*, ed., Colin Gordon (New York: Pantheon).
——, 1995, *Discipline and Punish. The Birth of the Prison*. Trans. Alan Sheridan (New York: Vintage Books, 1995).
Fox Keller, Evelyn, 1994, 'The Paradox of Scientific Subjectivity', in *Rethinking Objectivity*, ed., Alan Megill (Durham and London: Duke University Press) pp. 313–331.
Freedman, Barbara, 1991, *Staging the Gaze. Postmodernism, Psychoanalysis, and Shakespearean Comedy* (Ithaca and London: Cornell University Press).
Freud, Sigmund, 1961a, 'Three Essays on a Theory of Sexuality', *The Standard Edition of the Complete Psychological Works*, trans. James Strachey (London: Hogart), vol. 7, pp. 135–243.
——, 1961b, 'The Ego and the Id', *The Standard Edition of the Complete Psychological Works*, trans. James Strachey (London: Hogart), vol. 19.
Fried, Michael, 1968, 'Art and Objecthood', in *Minimal Art. A Critical Anthology*, ed., Gregory Battcock (New York: E.P. Dutton & Co) pp. 116–147.
——, 1980, *Absorption and Theatricality: Painting and Beholder in the Age of Diderot* (Berkeley and London: University of California Press).
——, 1987, 'Theories of Art after Minimalism and Pop', *Dia Art Foundation. Discussions in Contemporary Culture*, ed., Hal Foster (Seattle: Bay Press) pp. 55–58.
——, 1990, *Courbet's Realism* (Chicago: The University of Chicago Press).
Fuchs, Elinor, 1996, *The Death of Character: Perspectives on Theatre after Modernism* (Bloomington: Indiana University Press).
Gallop, Jane, 1985, *Reading Lacan* (Ithaca and London: Cornell University Press).
——, 1988, *Thinking Through the Body* (New York: Columbia University Press).
Garber, Majorie, 1992, *Vested Interests: Cross-Dressing and Cultural Anxiety* (New York and London: Routledge).
Gibson, James J., 1966, *The Senses Considered as Perceptual Systems* (London: George Allen and Unwin Ltd).
Gombrich, E. H., 1968 [1960], *Art and Illusion. A Study in the Psychology of Pictorial Representation* (London: Phaidon Press).
Greenberg, Clement, 1961, *Art and Culture* (Boston: Beacon Press).
Grosz, Elisabeth, 1990, *Jacques Lacan. A Feminist Introduction* (London and New York: Routledge).
——, 1994, *Volatile Bodies: Toward a Corporeal Feminism* (Bloomington and Indianapolis: Indiana University Press).
——, 1995, *Space, Time and Perversion. Essays on the Politics of Bodies*. (New York and London: Routledge).
Halkes, Petra, 2001, 'Aspiring to the Landscape. Investigations into the Meaning of Nature in Works by Wanda Koop, Stephen Hutchings, Susan Feindel and Eleanor Bond'. PhD dissertation published in 2006 as *Aspiring to the Landscape. On Painting and the Subject of Nature* (Toronto: Toronto University Press).
Jay, Martin, 1988, 'Scopic Regimes of Modernity', in *Vision and Visuality, Dia Art Foundation. Discussions in Contemporary Culture*, ed., Hal Foster (Seattle: Bay Press) pp. 2–23.

Jay, Martin, 1993, *Downcast Eyes: The Denigration of Vision in Twentieth-Century French Thought* (Berkeley, Los Angeles, London: University of California Press).

Jay, Martin, 1996, 'Vision in Context: Reflections and Refractions', *Vision in Context: Historical and Contemporary Perspectives on Sight*, eds, Teresa Brennan and Martin Jay. (New York and London: Routledge) pp. 1–11.

Kenner, Hugh, 1973, *A Reader's Guide to Samuel Beckett* (London: Thames and Hudson).

Kerkhoven, Marianne van, 1994, 'The actor', *Theaterschrift*, 7, 8–29.

——, 1998, 'Op zoek naar een actualiteit van Brecht. Brecht en het recente theater in Vlaanderen', *Het gevoel heeft verstand gekregen. Brecht in Nederland*, eds, Edmund Licher and Sjaak Onderlinden (Amsterdam: Theater Instituut Nederland and Aristos) pp. 101–115.

Kleist, Heinrich von, 1982, 'On the Puppet Theater', *An Abyss Deep Enough. Letters of Heinrich von Kleist with a Selection of Essays and Anecdotes*. Edited, translated and with an introduction by Philip B. Miller (New York: E. P. Dutton) pp. 211–216.

Koch, Gertrude, 1989, 'The Body's Shadow Realm', translated by Jan-Christopher Horak and Joyce Rheuban, *October*, 50 (Fall) 3–29.

Krauss, Rosalind, 1981, *Passages in Modern Sculpture* (Cambridge, MA and London: MIT Press).

——, 1987, 'Theories of Art after Minimalism and Pop', *Dia Art Foundation. Discussions in Contemporary Culture Number One*, ed., Hal Foster (Seattle: Bay Press, 1981) pp. 59–64.

——, 1993, *The Optical Unconscious* (Cambridge, MA and London: MIT Press).

Lacan, Jaques, 1977, *Ecrits: A Selection*. Trans. Alan Sheridan (London: Tavistock Publications).

——, 1981, *The Four Fundamental Concepts of Psychoanalysis*. Trans. Alan Sheridan (New York and London: W.W.Norton & Company).

Lakoff, George and Mark Johnson, 1980, *Metaphors We Live By* (Chicago: Chicago University Press).

Leder, Drew, 1990, *The Absent Body* (Chicago and London: University of Chicago Press).

Lehmann, Hans-Thies, 1997, 'From Logos to Landscape: Text in Contemporary Dramaturgy', *Performance Research*, 2 (1), 55–60.

——, 1999, *Postdramatisches Theater* (Frankfurt am Main: Verlag der Autoren).

Man, Paul de, 1979, 'Semiology and Rhetoric', in his *Allegories of Reading: Figural Language in Rousseau, Nietzsche, Rilke and Proust* (New Haven and London: Yale University Press) pp. 3–19.

——, 1984, 'Aesthetic Formalization: Kleist's Uber das Marionettentheater', in his *The Rhetoric of Romanticism* (New York: Columbia University Press) pp. 263–290.

Martin, John, 1939, *Introduction to the Dance* (New York: Dance Horizons Incorporated).

Melville, Stephen, 1996, 'Divisions of the Gaze, or, Remarks on the Color and Tenor of Contemporary "Theory"', in *Vision in Context: Historical and Contemporary Perspectives on Sight*, eds, Teresa Brennan and Martin Jay (New York and London: Routledge) pp. 101–116.

Merleau-Ponty, Maurice, 1968, *The Visible and the Invisible*. Trans. Alphonso Lingis (Evanston, IL: Northwestern University Press).

Mitchell, W. J. T., 1986, *Iconology: Image, Text, Ideology* (Chicago: University of Chicago Press).

Müller, Heiner, 1989, *Explosion of a Memory*. Trans. Carl Weber (New York: PAJ Publications).

Nöth, Wilfried, 1990, *Handbook of Semiotics* (Bloomington and Indianapolis: Indiana University Press).

Oosterling, Henk, 2001, 'Reflecting on the "Informe" and the "Inter": On Politico-Theatrical Conceptuality', in *Interakta 4. Performance, Transformance, Informance: New Concepts in Theatre*, eds, Henk Oosterling en Luk van den Dries (Rotterdam: Erasmus University Department of Philosophy) pp. 9–16.

Pavis, Patrice, 1998, *Dictionary of the Theatre: Terms, Concepts, and Analysis*. Trans. Christine Shantz (Toronto and Buffalo: University of Toronto Press).

Phelan, Peggy, 1993, *Unmarked. The Politics of Performance* (London and New York: Routledge).

Pontbriand, Chantal, 1982, 'The Eye Finds No Fixed Point on Which to Rest ... ', translated by C. R. Parsons, *Modern Drama*, 25, 154–162.

Rokem, Freddy, 2003, 'Where to Look? Constructions of the Spectator in the Modern Theatre', *MASKA*, XVIII (2–3), 14–20.

Rotman, Brian, 1987, *Signifying Nothing. The Semiotics of Zero* (Stanford, CA: Stanford University Press).

Said, Edward W., 1985, *Orientalism* (Harmondsworth, Middlesex: Penguin Books).

Sawday, Jonathan, 1995, *The Body Emblazoned. Dissection and the Human Body in Renaissance Culture* (London and New York: Routledge).

Schneider, Rebecca, 1997, *The Explicit Body in Performance* (London and New York: Routledge).

Silverman, Kaja, 1983, *The Subject of Semiotics* (New York and Oxford: Oxford University Press).

——, 1992, *Male Subjectivity at the Margins* (New York and London: Routledge).

——, 1996, *The Threshold of the Visible World* (New York and London: Routledge).

Stafford, Barbara Maria, 1991, *Body Criticism. Imaging the Unseen in Enlightenment Art and Medicine* (Cambridge, MA and London: MIT Press).

Szondi, Peter, 1963, *Theorie des Modernen Dramas (1880–1950)* (Frankfurt am Main: Suhrkamp Verlag).

Thacker, Eugene, 1998, 'Digital Anatomy and the Hypertexted Body' at: http://www.ctheory.net/articles.aspx?id=103

——, 2001, 'Lacerations: The Visible Human Project, Impossible Anatomies and the Loss of Corporeal Comprehension', in *Culture Machine*, 3, Virologies: Culture and Contamination at: http://culturemachine.tees.ac.uk/ frm_f1.htm

Tyler, Mike, 1997/1998, *Holoman: Digital Cadaver*. Unpublished text.

Wilshire, Bruce, 1982, *Role Playing and Identity. The Limits of Theatre as Metaphor* (Bloomington: Indiana University Press).

Worthen, W. B., 1998, 'Drama, Performativity, and Performance', PLMA Proceedings, 1093–1105.

Zonneveld, Loek, 2004, 'Simplicity, Beauty, Eloquence' *Holland Festival 2004* (Amsterdam: Holland Festival) pp. 166–168.

Index

Note: Page numbers in **bold** are references to figures.